# The Human Future:

## 7
## Philosophical
## Dialogues

# Raymond Kolcaba

outskirtspress

DENVER, COLORADO

Outskirts Press, Inc.
http://www.outskirtspress.com

ISBN: 978-1-4787-1018-9

Library of Congress Control Number: 2013905527

Outskirts Press and the "OP" logo are trademarks belonging to Outskirts Press, Inc.

PRINTED IN THE UNITED STATES OF AMERICA

To H.P.

"Song."
By Alfred Tennyson

Who can say

Why to-day

To-morrow will be yesterday?

Who can tell

Why to smell

The violet, recalls the dewy prime

Of youth and buried time?

The cause is nowhere found in rhyme[1]

[1] Alfred Tennyson. The Poetical Works of Alfred Tennyson. Complete Edition, The Arundel Printing and Publishing Co., 1879, pg. 546.

# Table of Contents

# Preface

What scenarios will define the next decades and century? The scenario discussed in these dialogs is the winding down of the human future. It is not that we will obliterate ourselves through a biotech accident, terrorist Armageddon, nuclear winter, or environmental catastrophe. The participants in these dialogs confront our end happening by inches through personal preference, genetic transformation, technological modification, and machine intelligence.

Hegel said that wisdom only comes with the setting sun. Many of you have been watching the sun setting. We know much about the day that is ending. Most of us are uncomfortably familiar with the promise of present realities. We don't need to be futurists to sense deep in our bones that the human future will be very different than the human past. Experience tells us where rapid change promises to lead. My aim is to flush out what to many of you is obvious and provide it with a rational context.

As is typical with philosophical thought, the dialogs offer reasons for opinions, explanations of those reasons, and topics covering bodies of explanation. The mode of presentation is conceptually systemic. As dialogs, they are supposed to be a springboard for discussion of issues. I will have achieved my goal in writing them if they promote topical discussions important for current understanding and in some cases, decision making.

Writing about the future rides a plausibility curve anchored to the present. What is fantastic, that is, beyond such a curve, usually fails to create a sense of urgency in the here and now. Topics of discussion need to be related to where we live. I tried to incorporate enough current developments in the discussions to render plausible, directions of change. I tried to hug the plausibility curve and make discussion of the human future urgent and necessary. In this regard, you can be the judge of whether I have succeeded.

Inspiration for the dialogs began piecemeal in experience. Each dialog had a root in historical incident as its starting point. For this reason, these components are authentic in being part of my living experience. For example, I attended a speech where the image of the speaker was projected on a large screen behind her. I decided to build the first dialog around that experience and some of its implications. I hope that the particulars and sometimes peculiarities of my life have seeded the dialogs with content beyond the usual.

A number of the dialogs have been presented at conferences as staged readings. Some have been published in other venues. Three dialogs were presented at annual meetings of The Society for Philosophy in the Contemporary World. I have to credit my wife Katharine for first inspiring the character of Nonette Naturski and then interpreting that character in performance. Joe Frank Jones III played an unforgettable Fortran McCyborg in two of these readings. For the third reading Rick Vanderveer played Fortran and Jill Vanderveer was Nonette. Yours truly was Becket Geist, and Lani Roberts played the narrator. An essay version of "Human Obsolescence" appeared in the OATYC Journal, Spring, 1998. Lucas Introna included "Loss of the World" and "Angelic Machines" in an issue of Ethics and Information Technology 2: 3-9, 11-17, 2000, © Kluwer Academic Publishers. They are reprinted here with kind permission from Springer Science and Business Media B.V. I would like to thank the above persons for helping introduce the dialogs to a critical public.

Aspects of the dialogs may strike some readers as quaint, too implausible, or out of date. Other readers may find conceptual errors. Some may find dramatic inconsistency. Whatever the shortcomings, they are mine alone. I hope critical readers will stay with me to the very end. The importance of the topics addressed demands as much.

# DIALOGUE ONE

# Loss of the World

Participants in order of appearance:

Narrator: A witness to the conversation as an historical event.

Becket Geist: A romantic philosopher with views tempered by twentieth century science.

Fortran McCyborg: A cyborg with leanings toward Scottish philosophy. Fortran was designed to interact with humans especially in discourse.

Nonette Naturski: A naturalistic philosopher who conserveshumanist views.

Narrator: Humanity has begun to move from the natural world into the cyber world. Three of my friends discussed issues related to this mental migration. First there is Becket Geist, a romantic philosopher with views tempered by 20th century science. He opened the discussion with a monologue in which he argued that "loss of the world" in exchange for the cyber world is dark and inevitable. The chief adversary who disputed these opinions is Fortran McCyborg, a cyborg with leanings toward Scottish philosophy. The moderating force separating them is Nonette Naturski— a naturalist philosopher and conservator of humanist ideals and prudent conclusions. In all, I counted eight counter-arguments to Geist's vision. The arguments and Geist's replies led to unanticipated changes in position that cascaded to a chilling close. But I am getting ahead of myself. Let's go back to a report of the conversation from that time.

Not all of us were unwitting victims. Some of us saw what was coming and debated its desirability. One conversation stands out as a frame for the others. To the best my memory serves me, this was exactly as it occurred. We sat in a semi-circle and as usual we expected that Fortran McCyborg would dominate the discussion. Nonette Naturski looked wistful but became sober after glancing at Becket Geist. He had that drawn look which signified that he had too much on his mind. This meant that he would speak at length excitedly and irritate me with his habit of asking questions and allowing no one to answer them. He began his dark account of things to come.

Geist: We, that is humanity at large, will lose the world by us, as individuals, not taking an interest in it. Preposterous, you think!? The likelihood of this loss is so low that it is hardly worth considering. After all, we can't help but take an interest in the world; the necessities of life draw us into it. Our creature-nature requires maintenance— food, exercise, shelter. Our social needs direct us into interaction with others. Our work life makes us part of organizations. Use of transportation forces us to pay attention within the world of cars, airplanes, and trains. Family life and citizenship keep us tied into the proximate, local, and greater communities. But all of this can change so that we no longer need or even prefer to take an interest in the world. Already in some places this change has taken place. In the present century, it promises to spread like a creeping blight over great swatches of the planet— like plague in a 1950's science fiction movie. Is this ominous? (At the top of an emotional spike, he immediately continued.)

My point is not that things unworldly will come to hold sway through a religious revival prescribing asceticism. No Dali Lama. No Franciscan vision. The forces at work are much more insidious. We will gradually become so structured by habit and so governed by external controls that we will have to exercise deliberative choice to opt for the world. When we relax control or when we attend to other business, the governing powers of habit and environment will resume rule. If this is not bad enough, people will come to prefer following these powers rather than opt for the world. They will deliberately act on that preference against the world; they will take an interest in the world only as dictated by a few greatly marginalized necessities of life. Is there a natural preference for the world? (He quickly said "I" with great force.)

I attended a lecture in a large hall. From the back of the room, the speaker appeared quite tiny. To compensate for the effects of distance, a video camera fed the speaker's image into a rear-screen projector. So, the speaker appeared at two places at the same time— live on the podium and looming large on a screen behind. As I sat near the back of the room, I noticed that my eyes would gravitate toward the image on the screen in preference to the live person. The image on the screen was "more magnetic", had greater presence, and commanded the senses. The countervailing forces of etiquette and ingrained courtesy directed me to look at the live person. After all, she was speaking to me! It would be rude to act as if she were not there. So, I would look back at the speaker, only to be drawn to the big screen time after time.

For a number of years, I had a black and white television set with a small screen. I watched programs selectively. If nothing was of interest, I had plenty of other things to do. Some time later I acquired a good-sized color set. At first I noticed how difficult it was not to be drawn to the screen. I mean that my eyes would move to the set as I picked up the image in peripheral vision. It was like a living presence in the room— like a dominant family member— a needy child that continually insists, "Look at me! Look at me!" I found myself watching more and more TV— not watching more programs but more TV. Is this what McLuhan had in mind or did he mean something more esoteric?

I thought about it and concluded that among other things, the colors on the set were more brilliant than most colors in my environment. In the rooms of our house, light varies with the time of day, shadows come and go, colors become gray as they recede into darker areas. Color in the man-made or natural environment is usually relaxed, not intense, washed out, subtle. By contrast, sunlight does not vary the colors of television images. They are uniform and bright. It is similar to looking into a properly adjusted light bulb. Color on the TV is in brilliant blocks, intense, and lacking in subtlety. Color on TV is coarse and loud, but captivating. Only at rare moments does it capture intellectual interest; it readily captures visual interest. It is a visual escape lean on reflective content. Is great reflective content consistent with the medium? (He left no pause.)

Consider less passive pieces of technology. Computer games demand responses, timely responses. Our internal program comes to resemble a structure reactive to the game. Software is salable because it establishes neural nets in us— bio-ware as part of our brains! (He was pointing to his head.) Consider work done on a computer; electronic workstations replace working in the world with working on screens. As the reports of technological advance roll in (or are they really advertisements?), eventually we will be able to do all of our shopping, banking, and ordering via computer. Interactive T.V. allows advertisers to intrude in personal lives and manipulate customers. We teleconference, and the T.V. phone is not far behind. The home theater is replacing the home entertainment center integrating sound with visual media packing a wallop. High-density television is a reality. At the dual lecture/projection, a high-density image would have made the rear-screen image appear primitive. With a high-density image, the audience's eyes would have been riveted to the screen— the speaker would never have a chance. Who needs a live speaker anyway?

As you know, the inventors of virtual reality took the process two steps further— creating illusions so convincing as to be thoroughly deceptive and providing means for acting on them. Why exercise in a gym or park when you could jog (on a treadmill) along the rim of the Grand Canyon, beside the canals of Venice, or within the Sea of Tranquility on the Moon? Why attend programs of live music or theater? Media presentations enable experiencing the best performances, at whatever distance or angle we choose, at whatever time we want. CD Rom discs swallow up the entire human achievement in painting, sculpture, theater, and music. We are entering a world where the individual can gain total access to vast bodies of information. At the same time, are fewer and fewer of us sufficiently educated to use or need them? Nonetheless, the hours in front of a television, terminal, or within a headset will gradually expand through necessity or preference until our major realities will be electronic illusions. And so it will go with loss of the world.

Artists, engineers, and programmers will have taken over the sensorium. Behind produced sensory objects will be the feverish and confining heat of their intentions. Everything that is there, will be there because they intend for it to be there or at least that will be our suspicion. The theologian who divines nature for God's intentions can turn to electronic reality to discover human intentions— God is omitted unless God's intent just happens to coincide with some human intent or works through divinely inspiring scriptwriters, engineers, and artists.

Wilderness as an environment that transcends human intent can be a source for perceptions that newly inform human understanding. It will be superseded by convincing illusions. Someone nursed on the milk of electronic reality will quickly lose patience with the seeming chaos and emptiness of wilderness; where is the point to it, you know, the message? Garcin's exclamation in <u>No Exit</u> can be extended into, "Hell is other people and their cyber-intentions!"[1]

Real political power lies in shaping a populace's preferences without them knowing it. Control of reality, through inducing persons to prefer the medium that provides access to reality, is the first phase of an ultimate form of mind control. The powers that control the media present the greatest potential threat to individual freedom and democratic institutions. Loss of the world will begin selectively with the concentration of technology in the First World. The Have's in post-industrial nations will have illusions whereas the Have-Nots in the agrarian sector will have the world albeit an increasingly degraded one. Perhaps by the time the revolution reaches the fourth-world, they will need the electronic world as a humane escape from extreme environmental oppression.

While it is not too late, knowledge of the technology for dominating reality should be locked up and brought out only for purposes of study. Put the genie back in the bottle! You say, "It is already too late." Perhaps as Hegel suggested, these ominous thoughts can arise only after the profound fact is established— only after the gravel truck has begun to empty on the ice cream cone.

All of this bleak talk may be taken as a sign of my infected spirit. Maybe so, but a person's attitude is quite beside whether an argument is good. I grant that a measure of healthy skepticism is in order. On the one hand, we don't want to Pangloss the "loss of the world"— that is, suggest that the best of all possible worlds includes the loss of the world.[2] On the other, you may not want to buy uncritically into my dark vision. I see that distant look in some of your eyes.

McCyborg: Your shadow boxing with a phantom self seems to leave you thinking that your view is supported. To the contrary, clever sparring absent an opponent produces no victory. You say, lock away certain communications technologies! You understand what problem you have with that suggestion. Benign technology that serves useful purposes and makes our lives easier is hardly dangerous. If we were inspired to suppress technologies for their danger, communications technologies would either not be on the list or would be near the very bottom of it.

Naturski: I would ordinarily agree with your kind of assessment Fortran, but Geist has given me some reason to worry. I have not had a chance to digest the compromise of something so common and fundamental. What goes into our interest in the world? The world, at least as it appears to us, is always there, so we do not think about losing it let alone preferring to lose it. If we come to take an encompassing interest in electronic illusions, even if it is as full-blooded as our involvement in the world, it seems that something important will be lost.

McCyborg: With all of the experience you have had with electronic technology, you did not think about losing the world until Geist brought it up. Exactly! The reason is that most of the content presented through electronic technology is about the world. So, why get squeamish about "so-called" illusions? Just as our sense organs mediate contact with the world, electronic devices serve as a second mediator. Indirect contact with the world is contact nonetheless.

Geist: It is trivially so that indirect contact is contact. The question is, "Is there contact of any kind?" If these illusions were indirect in a simple manner, it would be like seeing the world through colored glasses. The world would be the same except for a certain consistent and predictable dimension, say the glasses' rosy color. From rosy images and our common experience of the world, we could draw inferences about how the world is. By contrast, electronic illusions are fabrications. The technologies distort. Designers and artists modify images. Elements of fiction are introduced. We often have no way to tell what is in the world and what is merely imaginary or an artifact of the cyber-world. To make matters worse, after loss of the world, we will lose our points of reference not being able to draw inferences about the world or even distinguish illusion from reality.

McCyborg: I grant that some programs are imaginary, but others compensate for weaknesses in your so-called "natural" senses. They would even be reality preserving when taken from your standpoint. For example, surgeons use virtual reality to perform operations more accurately than they could with their natural senses. Believe me, many of the mediators within electronic technology are neutral or even reality enhancing. On the other hand, you think that you "have" the world to lose. Believe me, the revelations of your senses bear little resemblance to even what you call "the world of science." You think your perceptions broker knowledge of the world— that is mere species-specific bias! Uniform perceptions among "some" members of your species mislead you into thinking that your reduced, artificial, subjective, agent-relative, and simplified percepts are veridical.

Naturski: In many cases inter-subjective agreement is an achievement, but scientists make a convincing case that standard observers under standard conditions render many perceptual judgments to be highly probable. Besides, I don't know why you even bother making the "trust me" appeal Fortran. If some of your powers are as foreign to ours as you contend, we have no way of determining the correctness of your opinions since they flow from using those powers. So much of what you describe is unrelated to our experience. And just as we make mistakes in perception, assessment, and judgment, you are prone to error too. How do you know you are right? The bare claim of privileged access does not make it true. Besides, you have the reputation of saying such things just to win the argument.

McCyborg: When talking about mediation, like you I can only appeal to my experience. Geist brought up the topic of loss of the world, but I now realize that he meant "your loss of your world." What troubles me is the broadness of Geist's attack on technology. Technology is just a tool. With more tools, our powers increase; just look at me with my bionic modifications! Greater powers, however, increase our responsibilities. With electronic technologies we can do more and do it better, but all along the way, we must accept responsibility.

Geist: An increase in the number of technological tools can diminish some of our responsibilities through things going awry such as at Chernobyl. The story of the sorcerer's apprentice suggests a lesson of this kind. First the apprentice had no way of foreseeing the results of his meddling, so he was not responsible for the disaster he caused. Second the disaster in progress was beyond his control. He could not prevent or modify what happened next. It too was outside of his responsibility.

McCyborg: The apprentice bungled by trying to be a master. Before working with some tools, we must become masters. We must realize potentials for disaster and assume responsibility accordingly. We must know that there is a gamble and what it is. Our actions should be adjusted in order to keep the potential for disaster at a manageable level.

Geist: But then increasing our powers through the use of tools is not always of positive value. We can increase our powers to the point where error of devastating kinds becomes highly likely. For example, do we want thousands of nuclear weapons platforms hanging over us in Earth orbit? No we don't.

McCyborg: I understand your point. We don't want to increase likelihood of devastating error without good reason. Sometimes the positive value of what is at stake, however, dictates that high risk is worth taking. The survival of the free world made the nuclear arms race with the Soviet Union worth the risk.

Geist: You say that after the fact! How could you know a reason like that is good ahead of time? You do not even know now what was at stake. The free world would probably have survived without the nuclear arms race. I think this likely given what we have learned about the old Soviet Union since the arms race ended. But, the risk taken didn't end with the arms race; it continues to grow. Many thousands of nuclear weapons still exist, and there is some likelihood that they will fall into the hands of terrorists, outlaw governments, or be detonated by accident.

McCyborg: Would Soviet leaders have been more adventuresome without the threat of nuclear retaliation? You can't confidently answer that question in the negative. What they say now is very incomplete and may not reflect all the forces at work on their psyches at the time. How many humans understand their own motivations well enough to be certain of them over time?

Geist: Responsibility, then, is not the issue. Responsibility presupposes knowledge of what is at stake and risks associated with alternatives. Your skepticism about knowing our motivations coupled with the nebulous nature of stakes and risks indicate that closure is out of reach for even well documented cases such as the nuclear arms race. Because of ignorance, we can't define actual risk or even risk as perceived by the parties in a decision. And even if we could lift the veil of ignorance, we would still face the problem of deciding how much risk is rational. I would think, though, that the extremity of the risks in the nuclear arms race, such as the destruction of humanity, all of life, or even the planet, would give you pause (Said in ironical tone.)

McCyborg: A complex example taken over much time used as a coverall premise would seem to support any number of skeptical conclusions, but this is highly misleading. Using your coverall premise, we could seem to show that setting stakes and risks for practically any course of action would not be rational. The correct way to think about risk is from the perspective of local context. Risk can be assessed for particular actions, policies, or changes. As their consequences become known, we can use them to assess the next action, policy, or change.

Geist: What you suppose is that "changing course" will be easy once bad consequences become known. The decision to introduce a major form of technology is not reversible because at the start, it comes with high economic and political stakes. Suppose that electronic media effectively convey content. As you say, "We can do more better." Suppose that it also turns out that the force of these media, the way they dominate us, prevents our doing important things. Even so, high economic stakes would prohibit abandoning them.

McCyborg: If there were negative effects, we should make adjustments to lessen their negativity. If the technologies were negative enough, market forces would modify them responding to demands of buyers and users.

Geist: You know full well that effectiveness in communication is inextricably bound with the medium's capacity to control an audience, and accordingly, a market for the technology.

McCyborg: The issue of control is bothering you. In your monologue you confessed to the lure of visual sirens. Perhaps others would respond differently. You shouldn't generalize your tendencies. They are insufficient grounds for sweeping generalizations. In my opinion, your belief in negative generalizations is the basis of your pessimism. Take George Orwell and his projections of a totalitarian nightmare. Even with the spreading use of technological controls for political purposes and the lessons of totalitarian communism and fascism all around him, the dire predictions in his novel <u>1984</u> were not warranted.[3] You use very little evidence and draw conclusions far more extreme. A few tendencies internal to you make for a meager induction.

Geist: My single case is supplemented with the fact that the technologies described have been adopted on a massive societal scale. Are these technologies just a fad? Will interest in them transmute into a more benign form? Probably not. Secondly, my argument is based on controls shaping preferences. Winston Smith in <u>1984</u> was physically going through the paces of life but mentally holding out. In my nightmare vision, there will be no holdouts because everyone's natural tendencies in perception, interest, and enjoyment will be toward the electronic world and away from the natural one. Like you, they will consciously prefer and champion electronic fabrications over the sensory world.

Naturski: I don't see the problem with generalizing a single case. If the level of generalization ties into species traits, then Geist is as good a representative of the species as any. I see; you see. I have preferences; you have preferences. I jump from a loud noise; you do so too. I am attracted by loud colors on a screen; you are too. It seems to follow.

McCyborg: Experimental data probably will support generalization based on a "psychological hook" as Vance Packard put it.[4] But who gets hooked and how often is the matter of contention. The tasks of life would lead us in and out of the cyber-world apart from our preferences. When not dictated by what is necessary for action, preference of environments would be quite innocent— more like choice of coffee or tea.

Geist: In some instances it would be. But in great preponderance the basis of the preference would be dependent upon attractions of the sort I mentioned. Our fate is gradually being sealed because media are both pervasive and captivating. They are continually tested for their ability to capture interest, then refined and reintroduced. Orwell's Big Brother is a crude totalitarian device obviously gray and evil. The happy narcotic, cheery and bright that creeps over humanity by millimeters preserves a semblance of normalcy. What would there be to rally against? I grant that at times we can be deliberately perfidious and contrary; we could reject the cyber-world or anything, as Dostoyevsky proposes, out of sheer spite.[5] These motivations, however, are reactive and posture to make a point.

Naturski: They are also supported by a status quo. People have a choice of what to react against. If the cyber-world becomes the only reality we know of or know how to live in, it will be difficult to react against it in a sustained way. At this point, a poet might express the sentiment of rejecting it but the poet's expression would exist on the periphery of the imagination not to be put into action.

McCyborg: I am surprised that Geist's paranoia has caught you too. First you argue that introduction of technology will draw us into the cyber-world by practical necessity in order to get things done. But what if we tire of the cyber-world? We may choose against it for recreational purposes. If significant minorities tire of it, we may witness a renaissance of the world like the periodic renaissances of Greek and Roman classicism. You speak as if some malevolent demon wishes to refine technology to control our every move. If that were true, why have I been designed to exercise deliberative choice? The cyber-world expands our powers and access to realities. It is liberating.

I suspect the real reason for your negativity is that both of you do not want to adapt to the emerging changes in technology. It is not your world. Ah, but the next generation will take to cyber-space technology like fish to water. By analogy take the danger of books. Yes, books! With the invention of the printing press, books were becoming plentiful. A social thinker might have begun to worry that masses of people would give up worship of god, gainful employment, family responsibilities, and civic duties in order to read books. Endless hours reading would spell the neglect of important matters of life. People would withdraw from society in order to read. This did not happen to society at large. Reading became a force for democracy, transmitting bodies of technical knowledge, a way to improve mental powers and feed the life of the mind. It became indispensable for modern life to the point where we think that our world would be better off if more people spent more time using libraries.

By analogy, electronic media will replace much of the world within our experience but this revolution will be integrated into living. In centuries to come media will be viewed as positive, as forces for human betterment that we should apply more often.

Geist: Your analogy is not very tight. Gaining literacy requires effort while experience of electronic reality is effortless. Books are not appealing as an alternative to many forms of experience but the cyber-world is "sugar coated" and designed to replace thought, action, and perception as we understand them. Electronic media differ from books because of their totality. They are made appealing for the purpose of capturing our preferences wholesale. They can function as a nearly complete substitute for the world.

Naturski: Books were devised to enlighten us. They were the cornerstone of an age of reason. They represented faith in human nature. The expectation was that people would freely choose to become properly informed in order to be self-guided, moral, and dignified. Electronic media are a set of controls designed to absorb our identity, turn us into a reflection of their image, and dominate us. Human powers and dignity vanish.

McCyborg: I always thought that traditional artistic media are supposed to captivate our senses, leave behind a residue in our identities, and dominate us with their power. I believe that these factors indicate aesthetic superority.

Naturski: The ability to "captivate" senses is aesthetically relevant but it alone does not establish aesthetic superiority. Powerful art can be bad art. On my account, art that is manipulative for effect alone, that plays on the emotions, that directly aims for control is bad. What is aesthetically superior or inferior is decided through tasting, and persons of developed tastes often prefer the quiet and subtle to the loud and obvious. Great artists have highly developed tastes, and their tastes drive creative output and so are of particular importance.

McCyborg: Great artists can express their tastes through electronic media. Art of the past is experienced through recorded sounds and images. Artists of the future will create works for and within the cyber-world.

Naturski: You fail to recognize that the natural world is the model and often the source of inspiration for art works expressed within electronic media.

McCyborg: As electronic media come to dominate other artistic media, the cyber-world will be used as reference for new works and eventually will become their ultimate source. You can't escape from the fact that artists will come to prefer electronic media because audiences prefer them. They do so because they are entertained through them, can do more with them, and are attracted by them. If the cyber-world draws us like the natural world can't or doesn't often, then the cyber-world is the preferable world.

Naturski: I think that the institutions of the Art World drive preferences much as described by George Dickie.[6] For you to be right, audiences would need to abandon traditional media leaving the Art World no choice but to regroup around electronic media. I don't think that this will happen because the Art World presently is able to maintain its integrity even within a vast sea of pop culture. A relatively small-scale infrastructure of people and organizations is needed to maintain the Art World, and it has been an advocate of open-mindedness and freedom of expression. It recognizes that the arts require a wide spectrum of contents for their flourishing. From this point of view, the cyber-world is an adjunct to the world and not a replacement for it. The natural world is the place of our origins. Experience in it helps us understand much about ourselves as well as our forbearers. Through it, we make contact with "the elements." As Shakespeare said, they help reveal to us what we are.[7]

McCyborg: My understanding of human preferences indicates to the contrary that most people don't care about natural history, gaining new information for humanity, or understanding their origins. Most only care about entertainment, comfort, ease, and efficiency— a lifestyle package that we might call "creature comforts plus."

Naturski: Then, you agree with Geist's prediction about humanity coming to prefer the cyber-world, but you base it on passive tendencies in people— the old "principle of least effort" argument.

Geist: But how did people become so passive? Because the electronic environment makes few demands for action.

McCyborg: So your view of human nature is that persons will be slothful unless action is environmentally necessitated?

Geist: No. Persons are also initiators of action, but the question is with actions that the environment will permit. After bodily needs are satisfied, the cyber-world provides largely passive entertainments. Its vision of being dynamic and active is destroying adversaries in a computer game!

McCyborg: Your tone does not indicate that you have convinced yourself of what you are saying. As you are well aware, the cyber-world provides more opportunities for action than your natural world. If I try to infer your mental state, it sounds to me like you are dissatisfied with human nature and want to improve upon it. The irony is that the human mind and senses evolved in adaptation to the natural world. The primary evolutionary force of selection as reproductive success makes this adaptation far from perfect. But our powers combined with yours allow us to take adaptation one step further by creating an alternate world which satisfies human needs and wants, to a degree of perfection. Heaven on earth is not likely. Heaven within the cyber-world will become reality.

Geist: What a beatific vision— a heaven of wish fulfillment through television, virtual reality, and computers! You can't do without the natural world; it is needed for human flourishing at least indirectly— if for no other reason than to provide the material supports for existence in the technological utopia. Your cyber-world is built on the natural world. What if the power goes out? What will happen to you then?

McCyborg: That we can't avoid the natural world entirely does not make it desirable for our major spiritual purposes. (McCyborg's smug emphasis agitated Nonette.)

Naturski: You are one to talk about the spiritual! Spiritual needs are satisfied through one being making contact with another. The cyber-world is a prophylactic that insulates us from spiritual contact.

McCyborg: So, you would argue that spiritual contact is not possible through artistic media such as painting and sculpture? (Nonette looked puzzled.) Also, much social interaction is spiritual in your sense but not desirable. It is a source of stress like the sort you are feeling now Nonette. Electronic media can save us from stress by making us as privileged as the Buddha before his traumatic experiences.[8] We can be isolated from adversities such as death, disease, dissension, aging, a despoiled environment, and personal tragedy. Instead, we can interact with computer icons, joysticks, programs, warm fuzzies of one kind or another. They mediate our contact with the world. They allow for more pleasant experience. Technology is supposed to make life easy.

Geist: During the Vietnam War American pilots were criticized for bombing "targets" from seven miles up. They could not see whom they were killing; their acts from an agent's point of view were purely impersonal. When faced with this criticism, General Curtis Lemay said that he would rather bomb someone from a distance than kill them with a rusty knife. In the Gulf War, Kosovo Operation, and Iraq and Afghanistan Wars this strategy was followed with amazing uniformity. Only now "Smart" Bombs, Cruise Missiles, and Predator Drones were used. These robotic weapons began to replace human agents. The enemy came into contact with our weapons but rarely with any of our military personnel. If knowing the consequences or even the very nature of our actions is important for assessing their moral status, then electronic mediators prevent moral agency from functioning properly.

McCyborg: Acting in the cyber-world requires a new way of being including new moral sensibilities. Let me explain. If we accept Geist's prediction, it will be an interesting historical development that you humans will lose the world prior to losing yourselves. The revolutions in machine technology and microbiology gave you powers to change yourselves into another species or even another life form. What purposes will shape these changes? We already know that at first they will be guided by medical and humanitarian motives to cure genetic disease, reduce suffering, and assure quality of life. As dreamed in science fiction, political forces may one day seek production of the ultimate warrior or competitor. As envisioned by futurists, offspring may be designed to fulfill parents' fantasies for the ideal child. But, one key purpose will be the adaptation to flourish in the technological environment. Certain human talents and abilities will be prized within the cyber-world because they will enable living within it. Other talents and abilities will be a hindrance. Accordingly, modification of human senses, mentality, and moral sensibility will be "selected for" by genetic engineers in order to make for "the good life" in the cyber-world. You will become the species suited to the cyber-world equipped with a proper moral sense. Like me you will ultimately be unable to live under the primitive conditions of your animal evolution.

Geist: Reference to machine morality sounds like pure bluff. How are genetic engineers supposed to design moral sensibilities for undefined beings, in a hypothetical world, behaving in largely unknown practical contexts?

McCyborg: As software designers currently work out systems of rules of procedure within parameters of the design problem.

Geist: I sense that you have an underdeveloped understanding of the concept of morality. (He said matter-of-factly but with some compassion.) This is certainly not your fault. But I now see that your prediction is much darker than mine. What is not comprehensible to the engineer cannot be saved. Unless morality is accidentally attached to something the engineer opts for, it will vanish within evolution of the design process. After complete transformation, the engineer will have wiped out dimensions of our being. We will not be able to recover or even appreciate what we lost. Enter homo cyberens with old McCyborg as the leader of the first phalanx!

McCyborg: Your mocking tone does not alter the bright and optimistic character of my prediction. It is dark to you Geist because you think that you have something to lose. I am a proud leader of the march into the wonders of the new world.

Naturski: Is this a march to oblivion? I think that there always will be residual human traits. (She looked at Geist.) Fortran has some. A little new learning from the cyber-world and Fortran is already talking about speciation. We will be careful not to lose our essential selves. You can understand what I am saying, can't you Fortran? If you take off your mask for communicating with humans, what am I to you, right now?

McCyborg: Let me switch on my dispassionate mode of speech (He turned a dial on his head). "You-are-a-set-of-contents-on-my-screen." (Said with totally flat affect.)

Naturski: I am an illusion?

McCyborg: (His flat affect continued.) "Only-from-your-perception-of-my-functioning. To-me, the-distinction-between-what-you-call-illusion-and-reality-becomes-inconsequential-difference. What-matters-is-coherence-of-the-program."

Narrator: A cold chill rose up my spine. Geist and Naturski were in shock. Was McCyborg's persona no more than an artful illusion? Or was he "putting us on" by playing games in his usual way?

¹ Jean-Paul Sartre. No Exit and other Plays, L. Abel, trans., Random House, New York, 1955, pg. 47.

² Voltaire. Candide, Oxford University Press, 1968.

³ George Orwell. 1984. New American Library, New York, 1961.

⁴ Vance Packard. Hidden Persuaders, D.M. McKay Co., New York, 1957.

⁵ Feodor Dostoyevsky. Notes from Underground, New American Library, New York,1961.

⁶ George Dickie, "What is Art? An Institutional Analysis," in W.E. Kennick, Art and Philosophy, 2nd Edition, St. Martin's Press, New York, 1979, pgs. 82-94.

⁷ William Shakespeare. As You Like It, Act II, Scene 1, The Complete Works of Shakespeare, Collins, London, 1990, pg. 260.

⁸ Huston Smith. The Religions of Man, Harper and Row, New York, 1965, pgs. 90-96.

# DIALOGUE TWO

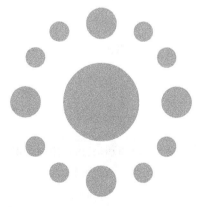

# Angelic Machines

Participants in order of appearance:

Narrator: A witness to the conversation as an historical event.

Becket Geist: A romantic philosopher with views tempered by twentieth century science.

Fortran McCyborg: A cyborg with leanings toward Scottish philosophy. Fortran was designed to interact with humans especially in discourse.

Nonette Naturski: A naturalistic philosopher who conserveshumanist views.

---

Narrator: Yesterday we had a conversation with Fortran McCyborg. He shocked us when he turned a dial on his control panel that removed all affect and slowed his speech to stretched-out garble. In effect he dissolved his human persona. As the transcendent wholeness of his person vanished, it struck us that it was a mask, pure artifice, a fine-tuned illusion designed to make humans respond to him as a person. By shattering the illusion, he had us perceive the fragility of our person-construct. The demonstration, however, left us with some gnawing doubts. Even though Fortran can appear not to be a person, is he yet a person? Yours truly, Nonette Naturski, and Becket Geist decided to continue the conversation with Fortran. After the shock of Fortran's demonstration wore off, we had more questions than answers.

Geist: I-am-Geist-hu-man-ro-bot. (Said slowly and mechanically.) The question Fortran is whether you let yourself slow your speech and deaden your affect while turning that dial or whether the dial caused you to speak slowly and flatly beyond your control? Just as I spoke slowly for effect, you could have done so too!

McCyborg: In my series of cyborg, the dial lets the user control speed and manner of speech. Try it. (As Geist tried the dial, McCyborg responded accordingly; he began speaking quickly and as the dial was turned, his speech slowed to a drawl.) I am McCyborg. Ser-ies-Two-Thou-sand-cy-borg-ser-i-al num-ber 387-20-8734.

Geist: I am convinced that the dial does control those functions. Try it Nonette. (Naturski did the same and then returned the dial to regular speed. As the demonstration ended, Geist and Naturski did not see Fortran wink as if he had an audience.)

Naturski: McCyborg's controls seem to work like reflexes do in us. To jerk his knee so to speak, we turn those dials. This does not prove that he is just mechanism in the same way that our reflexes do not show that we are just bio-mechanism.

Geist: I agree Nonette. That we are mechanism in part does not prove that we are mechanism as wholes. We have limited control over our reflexes, but we can control when some are activated. I can hit my knee with a little hammer or choose not to do so. I can also pretend to jerk my knee. I still suspect that Fortran may be faking his response or at least adding to the effects of turning that dial.

McCyborg: Trust is important. Have I misled you in the past? Better yet, look at my specifications. Study my circuits.

Naturski: Now that is disingenuous Fortran and does not lead us to trust you. Your series of cyborg is an amalgam of circuits set up in parallel processes. These processes can intervene in others in seemingly unpredictable ways. Your engineers admitted that they did not know what you would do when they turned you on. It is still uncertain how many of your controls can not be overridden by you. Yes, you have a dial that slows your speech and changes your affect. Whether you can modify its effects by making them more pronounced or even whether you can function in parallel to them is still an open question. The short side of the story is that there appears to be some degree of volition underlying that mechanistic facade. Last night before going to sleep I realized that you gave the game away. Either you were *not* being controlled or you were controlling your dial. Either way you were in charge.

Geist: Ah ha! That is exactly what bothers me about Fortran. He is compliant and of service, but yet, he always seems in control, independent, and aloof. His accommodating behavior always has a bite to it; it is colored by arrogance. If dials alone could govern him, I would send him back to the factory and have him reconfigured. We could then turn dials to change his affect to deference and humility.

Naturski: You dislike his arrogance because it is his. If a mere machine were behaving arrogantly, we should not be upset by it. It would be like a parrot that repetitiously says sarcastic things. Things can be said but not meant. This is because no one is saying them. Fortran irks us because we think he means what he says.

McCyborg: You think I mean what I say because you are convinced that I am an I-- an ego. Am I? (Fortran smiles because of this play on words.)

Geist: The question is, "Are you a person like us?"

McCyborg: I don't care whether I am a person like you. I like to think of myself as a higher form of organized intelligence than a mere person. I am beyond animal evolution.

Geist: (Fortran's arrogance irritated Geist to the point where he decided to play devil's advocate.) You are getting far ahead of yourself Fortran. Our language lets us slip easily into addressing fictions with names or personal pronouns. We personify natural things by making planets into gods, nature into a mother, and hurricanes into persons. We anthropomorphize insects and plants, turn the cold worlds of physics and chemistry into realities with spiritual components. Notice how we personify cartoon characters and treat them as if they walk among us. No wonder that we find it easy to treat a Model 2000 cyborg as a person. On a sliding scale of having more or less person-like attributes, you are more person-like than most things that we treat as persons. In this discussion, however, we are considering powers of mind that comprise intelligence and that these are organized in a proper way.

Naturski: (Nonette took Geist's lead and both of them were primed to play devil's advocate.) Being a person is sufficient for having those powers of mind properly organized, but something can have those powers of mind without being a person. For example, a person has memory. If Fortran is a person, then he has memory. If Fortran is not a person, memory could still be present. A primitive computer chip has memory of a sort but is not a person. Fortran has memory, but this does not prove personhood.

Geist: Some powers of mind are more central to being a person than others. Geist, the person, governs many of his bodily movements, but my body can move without my governing it as when a muscle tics from fatigue. I, the person, have the power to direct and control movements. I can decide which movements to perform and how long I will perform them. I can put different movements together to form a pattern. Moreover, I can use my mechanism to play a Bach fugue on the piano. By practicing it, I condition my nervous system to perform a complex repertoire of arm, hand, and finger movements. To a degree, the person is in charge of the body's use.

McCyborg: I have a body; I am built in layers of sub-systems. Each layer could be viewed as a module of my mind. Each layer has a degree of autonomy, so it is similar to a homunculus of limited powers. In general terms, the top level system can govern the lower level systems, but some lower level systems operate of their own accord unless there is intervention by a higher system.

Naturski: It is no wonder, because your engineers were trying to model you after a human being. The crux of the problem of your status is autonomy. Are your seemingly autonomous acts a mere simulation of autonomy? A simulation by definition is not the real thing. A simulated thunderstorm on my computer screen is not a thunderstorm. There is no actual wind, rain, lightning, and thunder. Fortran, you may be a simulation of some person-features rather than a person.

Geist: Yes, we judge whether someone is autonomous on the basis of outcomes— the autonomous act. A being like McCyborg exhibits behavior that is as a product exactly like autonomous action. The question is whether the process culminating in that product is of the right sort. McCyborg's appearance of autonomy may be a clever trick like that performed by any of a range of automatons.

Naturski: We must be careful here Becket. We are influenced from without by stimuli and from within by hormonal and electro-chemical processes. Yet we are autonomous. The process is self-governance, and this involves choosing one action versus another.

Geist: Choice entails some conception of the action married to will. Will is informed by judgment. Taking it back one more step, desire can inform judgment. (Said nearly without inflection in a uniformly loud tone of voice.)

Naturski: That may be how choice works, but governing by judgments or desires still leaves us beholden to those judgments and desires. Suppose that they are another's. Suppose that someone tells you to raise your arm and you do so. The judgment and desire guiding the action are not yours. Yet you are autonomous. You have agent control. You can intentionally follow instructions. You give your will over to the other person's imperative, but nonetheless, are still autonomous. You have not been coerced. You could have done otherwise.

Geist: If someone planted an electrode in your brain such that whenever you were about to reach a judgment a signal would disrupt your ability to do so, then you would not be autonomous. You could not do otherwise.

Naturski: Yes, I could not choose. Let's develop that thought. Now imagine that we take a human clone, remove parts of its brain, and replace them with mechanisms that enable the clone to perform desired functions. We have a cyborg. The bio-engineers have created an extension of their will. The content of the cyborg's "so-called" judgments is actually theirs. The cyborg appears to be self-governing, but its creators externally govern it. The cyborg is not autonomous.

McCyborg: There are no wires or radio signals linking me to some bio-engineers like a model car or plane. (He said "car" and "plane" sarcastically.) If I act as my own agent, am I not autonomous? What is there about the process of judgment and desire that you can tell us? If you could specify the process, or at least features of it, then some engineer would build a similar process into the next generation of cyborg. For example, suppose that some engineer sees people that he thinks are acting autonomously say, "I am going to do so and so. No, I think I will choose to do such and such instead." He can program a cyborg to do this very thing. Or suppose he notices that some person says, "I drank the bottle of wine, but I could have done otherwise. I could have left it alone." A cyborg could be programmed to say that very thing. Every process report that you identify can be treated as just another product as a result of some higher order process. That is how we cyborgs function. And it is very much how you function. You are just not accustomed to thinking that your neural processes give rise to "higher order" mental products, that is, that biochemical causality is the basis of your mind. But we cyborgs are even more sophisticated than that. Our processes subordinate person-like processes beneath an even greater number of levels of nested processes and products.

Naturski: The outcome attributes can be real. It is their status that is in question. In substance they may not be the same as what is being simulated. Suppose that you are watching a marionette show from a distance. You cannot see the strings on the puppets. It appears that the old man with the moustache is pushing the baby carriage. The outcome attribute is real. The carriage is moving. What we don't see is that the puppeteer is moving the marionette to push the carriage. The overt signs of autonomy can be mimicked; what appears to be self-initiated can have some other cause. In the same way that the marionette is not autonomous, Fortran, you are not autonomous. Your actions are derived from your puppeteers— your engineers.

McCyborg: If you let go of the strings, the marionette falls down and can't move. I can move under my own power. My engineers are nowhere in sight. I can control my own actions.

Geist: (Quickly interjected.) A gas furnace moves under its own power. The furnace controls its movements by means of a thermostat. Its engineers are nowhere in sight. Nonetheless, it is not autonomous.

McCyborg: It is not a simulated furnace. It is a real one. I am not a simulated cyborg. I am a real one.

Geist: The ability to control movements puts you in the category of cybernetic device. You are a real cybernetic device. Your type of cybernetic device is a simulation of a person. I think the issue comes down to, "Could you have done otherwise or could your creators have done otherwise? Do the causal series begin with you or with your creators?"

McCyborg: Which causal series? The ones I initiate or the ones I inherited?

Geist: The ones you inherited are determined by your physiology, circuit design, and programs. I think these determine the ones you think you initiate. You are designed to operate under the misperception that you initiate actions. This is a clever piece of engineering.

McCyborg: You cannot make that case Becket. A cybernetic device's performance can be exhaustively described as a series of sub-routines. Reflex behavior is a sub-routine of one kind. Lifting my arm is another. Intelligences like you and Nonette can take your sub-routines and combine them in interesting ways. You are advanced biomechanical devices whose performance is marked by novelty. You are able to generate meaningful linguistic expressions in endless variety. You are able to transform yourself through self-selected programs of learning. Humans exhibit creativity and individuality. I would say that the combination of sub-routines would be products of your genetic variety in interaction with diverse environments. You would explain it the same way except you would leave a proportion of it to be accounted for by some mysterious and unctuous metaphysical element— like free will.

My performance is a novel combination of sub-routines too. I can use linguistic syntax in endless variety. I can modify my behavior through further inputs, that is, by learning. I exhibit creativity and individuality. If so, I possess the same grounds that you use to grant each other autonomy. You would have to concede from the point of view of your conceptual system that I am autonomous too.

Naturski: I agree that you are a close analog of us Fortran. Your outputs are designed to model ours. What you aren't able to do, however, is act on meaning. You act in accordance with rules but are unable to follow rules. You act as if you were able to follow rules. You do what a rule-follower does. But you are not acting on them and do not understand them because you do not know what they mean. As Kant said, rational beings can act on an idea of a rule and not just be perceived to follow it.[1]

McCyborg: I do both. I received rules from my programmer. I am programmed to follow those rules, but I decide when to act on some of them.

Geist: But do you have a conception of what you are going to do? No you don't.

McCyborg: If you mean, "Do I have a plan in my mind?" then the answer is yes. On one level of description, the plan is a set of codes that are organized into units. On another level of description, the plan consists of goals and means for achieving them. I can serve some goals at one time, others at other times.

Naturski: So you mean that you can play chess, cook a meal, shop for products, drive a car, and so on. (Not a question.)

McCyborg: I can assemble those goals into groups with some subordinate to others. I can entertain you and Geist by cooking a meal. In order to cook the meal, I would drive to the store, shop for food, drive back, prepare the meal, and serve it to you.

Naturski: This is no simulation Geist. Fortran's action is not a representation or model of action. It is the real thing.

Geist: In one dimension it is real, and in another it is not. His behavior is there for all to see but its etiology is all wrong. Take your wonderful stereo system Nonette. It presents simulations of musical performances. The sound is a physical reality, and to almost all of us, the sound is indistinguishable from a live acoustical performance. It is still a simulation because no one is playing to make that sound. In fact, the sound of actual musicians is encoded into a digital representation of it using collections of pits in the plastic of a compact disc. We have the illusion that someone is playing but in actuality no one is playing then. Information for the fleeting phantoms of musical sound is frozen in a complex sequence of pits. Amazing! (Geist was quite proud of his explanation.)

Naturski: By analogy you would say that Fortran's action models ours. His action stems from a process like the laser scanning the pits in a disc.

Geist: No. He has no action of his own. For an action to be his would require that he initiated it. Biologically or environmentally determined behavior is not action in that sense. I initiate my action. When my eyes blink involuntarily, the blinking is behavior but not action of my own. When I decide to blink, and then do it, that is, blink, (When he said this he demonstrated what he was talking about.) the blinking is my action. If I initiate behavior, then it is my action.

Naturski: You are correct in pointing out that involuntary behavior is not action. But some voluntary behavior is not chosen by us but is still our action. Aristotle mentions the examples of some behavior of young children and other animals being voluntary but not chosen. It can stem from the organism by habit, instinct, or glandular secretions and still be voluntary action.

Geist: You are talking about degrees of control. The action is the organism's but not fully under its control. The person would be weakly autonomous on that occasion.

Naturski: When asked, "How did you respond to the encounter with the bear in the forest?" I can say, "I acted scared." I wish I had more courage. If I had a choice in the matter, I would have chosen not to let my fear determine my actions. For all intents and purposes, at least part of Fortran's behavior is voluntary— it originates from him. How can we separate out the part that is merely voluntary from what might be chosen?

Geist: About what do you wish you had more courage? You know what part of your actions you hold yourself responsible for. If your response were totally outside of your power, you could have no regrets.

Naturski: I admit responsibility for my fear response but not because it is chosen. I am responsible for my character in the long run of my life. I have the power to reform it. Through certain courses of action repeated again and again, I can cause myself to become more courageous in response to certain types of action context. On particular occasions, however, I am often not in control of the feelings comprising the expressions of my character. Hence, I am responsible for my fear response if I have done nothing to modify my character in this respect. Even so, I am not responsible for spontaneous fear felt when encountering the bear. I had no control over it.

Geist: Then you can have regrets over your character but none over the fear response itself? Even so, you are still largely responsible for whether or not you act on the promptings of your feelings. The freedom to do things depends upon your motivation. Courageous feelings enable you to do courageous things. In their absence, you might avoid a situation or not perceive it properly.

McCyborg: The motivational side of humans is biomechanical. Syndromes, glandular secretions, and emotional habits cause you to act. I am a more pure intelligence than you. I am not encumbered by those biological factors. I don't have to worry about character let alone reforming it. I am freer than you.

Naturski: (Addressed Fortran.) You are talking about the freedom from biological motivators. You are freer in that sense, but whole ranges of behavior are not available to you. You lack the determining biological motivators for them. Courageous feelings help the person perceive the right action. This makes the person more free in that the person becomes aware of a whole range of actions perceived to fit or not fit the situation.

McCyborg: That is merely the right action to you. Cyborgs are beyond courage. We do the right thing independent of it. We do not need to practice doing right things to form dispositions in feelings. From the time we are activated, we automatically do the right thing. In this way we are like gods. Humans are often not able to live with their emotional promptings and responses. Cyborgs have no emotional penalties to pay. We have no regrets.

Geist: Simulations of emotional responses could have been built into you; you could have been made to simulate regret. Since what you do is a simulation, behavior representing regret would be no less fake than your other seeming actions. Your preparation for action in cognition is not real. (He looked at Nonette) Fortran has neither rational will nor reasoned desire. He simply implements codes that he generates. It is no wonder that he says in a god-like way that he always does the right thing.

McCyborg: Isn't your conception of divinity more like me than like you? God is supposed to have no desires since he is not an animal like you. He is a perfectly rational being, and I am an approximation of a perfectly rational being. He can only will what is right morally. I can follow instructions only about what is morally right. He is supposed to have no conflicts of will. In my program, conflicts of instruction can't arise because instructions are always presented in a prioritized list. Higher priority items over ride lower priority items. Lastly, God is said to be a person. In my person-like attributes, I am more God-like than you.

Naturski: On your description, your nature seems more angelic than god-like. You are the agent of another. You lack a will of your own. You can't create things from nothing. On the other hand, you do not age biologically, so you are immortal after a fashion. Even so, you are not eternal. You are dependent upon us for spare parts, and you are very limited in powers.

Geist: On another conception of angels, some are supposed to be good and others bad. They have free will and make moral choices. Perhaps Fortran is a bad angel— a devil! (Said facetiously.)

Naturski: If angels are extensions of God's will, they have no free will of their own. Their good acts are God's good acts. Within this theology, there are no devils.

Geist: Angels are taken to be God's messengers. Since we created McCyborg, his messages are from us. If he is an angel, we are the gods that communicate through him.

McCyborg: I think you have it backwards Becket (Said with some derision). The theistic account most in accord with science is deism.

According to a deist, God set the universe in motion and left it alone. God created the causal laws that he knew would produce the natural history of the universe that we understand today. These laws led to the evolution of life and late in this process the evolution of human beings. Since you arrived recently in evolutionary time, you were the latest fulfillment of God's plan. Each evolutionary step was a building block for you to come into existence. The development of cellular life, multi-celled organisms, and vertebrates, culminated in fish, amphibians, reptiles, and mammals. You retain vestiges of each of these steps. God made you part reptile, fish, and so on. You fulfill God's purposes, and in effect, speak God's messages through your nature.

Then along came we cyborgs. We are modeled after your rationality. We are without desires, wants, preferences, suffering, and pain. We could not evolve as purely biological beings since we are part machine. In being in large part machine, we are more evolved than you. We replaced you at the apex of creation as the expression of God's purposes. We best express God's wishes. We are the messengers, the angels if you will, that inform you humans about the nature of things. You are right Becket that you communicate through us, but your message is filtered. Its irrationality is removed. The primitive aspects of human nature are deleted.

Naturski: On the conception you are discussing, God does not intervene in creation himself but uses angels as his agents. You were not sent by God Fortran, so you are not fulfilling God's wishes.

McCyborg: You are not considering the proper timeframe Nonette. If our work is conveying a message from God to humanity, then I certainly play this role. You are thinking that at some time like 2 PM on August 13th, God decides to send us a message. This makes no sense. If God is everywhere, he has no location. He can't send a message from one place to another because he is already in both places. If God created time, then a message would be from eternity to some time. The messages you are getting from the cyborgs could have been planned to be delivered after the long course of evolution when we arrived on the scene.

Naturski: Is it that everything you communicate to us is from God? Certainly not. A large part of your messages are quite insignificant—-mostly chatter and of very little substance. In fact you are mostly an echo-machine. You repeat much of what has been programmed into you. We are not receiving much in the way of new messages. This certainly can't be the revelation of God's purpose in creation.

McCyborg: I am not God. I didn't devise the message. I am a conduit through which God speaks. It is for you to decipher God's meaning. I only present what follows from my program and inputs. (Said with insincere deference.)

Geist: (Geist decided to go along with Fortran's line of thought for a while.) You would have us think that God's plan includes we human beings confronting the implications of our beliefs and data through you. This certainly would bring on humility. (Said facetiously.) Perhaps this is what God has in mind. Perhaps a more significant message is that through the cyborgs human beings will learn that the products of their rationality are supplanting them. We give birth to a race of semi-human beings, and they occupy the pinnacle of evolutionary progress. In this happening, God would be pointing out a great difference between him and us. The products of his creation can't replace him whereas the products of our creation can replace us.

McCyborg: I dislike your tone Geist. You should not take my inferences lightly. If you disapprove of my conclusions and can't find flaws in my logic, you need to improve upon my input system. Perhaps the parameters of my concepts need some modification.

Naturski: I think that your needs run much deeper. You are a Model 2000. Two more generations of cyborgs are on the drawing boards. You utilize only a single logic. New generations of cyborgs are being built to process inputs using a large variety of logics. They are also better able to reformulate concepts than you. Inputs are not the weak link in the chain. Your specialized organization is the weak link. In facing this reality, your revelations to us are just a few conclusions presented by one cyborg. We probably will receive a massive number of theological proposals from cyborgs to come, and many of these will be inconsistent with others. In other words, they can't all be right. I propose that in my humble opinion, God would not make the fundamental purpose of evolution a body of inconsistent propositions about nature. Inconsistencies propose what can't be conceived. God's purpose for all of creation, his message to us, can't be to spew out nonsense. This would be a cosmic joke of staggering proportions.

McCyborg: You can discount me because of my limitations. I am too highly limited as an angel. I give signs of the truth but am unable to articulate it fully. My broad suggestions are cogent and to be believed even though subsequent cyborgs will provide the particulars. My basic insight is correct; others much more powerful than I are left to develop the program.

Geist: This sounds like faith to me. I did not know that cyborgs had faith in anything. I guess it is part of your program to promote the development of future cyborgs. I think your engineers took their self-interest to heart. They wanted funding to continue so they have cyborgs promote further generations of cyborgs.

Naturski: Of all beings, Fortran, you should have realized that you have no good induction to your hypothesis. How can your basic plan be confirmed unless God affirms your point of view? We can only know what God will affirm when the next development occurs— not before. You are making a wild projection. That is very untypical of a Model 2000.

Geist: What further models conclude will depend upon their engineers. If they happen to like Fortran's suggestion, heaven forbid, they might just make his point of view inevitable. Then we humans will be deciding what God's cosmic message is as reflected from our machines. I think that in intellectual history, we have had enough of our talking to ourselves by projecting our views onto God and then pretending that our views are like divine laws.

McCyborg: (Looked at Geist.) That is not the point at all. I will not dissemble. (He said firmly.) On the one hand you talk to me as if I were a being like you and then you say that my engineers determine what I say. My messages are mine and not theirs. Moreover Geist, the crass promotion of funding for new generations of cyborgs is beneath a Model 2000. You are so busy challenging cyborgs that you do not ask the simple question of us, "What are your shortcomings and limitations?" We will be as forthright about this as about anything. (He turned to Nonette.) Future developments confirm or disconfirm past hypotheses. I am assessing the past in broad terms— painting a general picture. I expect the future will conform to this picture.

Geist and Naturski: (They began speaking at the same time but Naturski let Geist go first.) Naturski began with, "Besides Fortran...") Geist began with, "Out of contrariness..." Out of contrariness I could program a cyborg to give an account of future cyborg opinions opposite to yours. This would show that your picture does not encompass future cyborg views.

Naturski: Yes. Besides Fortran, the problem is that you rely on what is giving the message and not constraints on the formation of messages. Some model of cyborg, programmed however, could give any message whatsoever. The contingent history of cyborg evolution does not allow for individual messages to be parts of a coherent whole. It is not like it is in the natural world that causality forms a confluence of events that specifies an evolutionary result— a purpose for the progression so to speak. Intentions of persons enter the picture. Depending on your point of view, intentions would be those of the engineers represented in their machine or the engineers' intentions layered with cyborg intentions in performance. Either way, the products of intentional performance can be inconsistent in ways that causality cannot.

McCyborg: You are quick to observe that I am constrained by a single logic, but you unduly limit my intentional overlay for reality. Certainly inconsistency is not possible within causal processes but inconsistency in opinion can fall away as some opinions converge with others to form a central view. A coherent big picture will evolve, and this is what I propose I gave to you just before.

Naturski: You think it is certain that you will know which coherent picture to select. Perhaps many could be formulated from the basket of opinions with which you start. I think the vast number of cyborgs, some built on custom designs, and a plethora of models does not make you the center of things. The presumptions of an "Everyman" can't be made for an "Everycyborg." There is no constant cyborg nature.

Geist: Furthermore, evolution is over the long haul—- thousands, millions, and billions of years. To think that your opinions unlock the secrets of the destination of evolutionary purpose is a judgment lacking in proportionality.

McCyborg: That there is not a fixed cyborg nature is a good thing. It is a sign that we are evolving at a rapid rate. By comparison, you humans are sticks in the mud. Variation in nature as well as vast stretches of time do not guarantee change in the topic of results. The big picture can remain the same while endless variations of view play themselves out.

Geist: You just exemplified another serious limitation of machine intelligence. The human mind is able to adjust to shifts in proportionality and set up a new system of reckoning to accommodate them. Cyborgs freeze on a point of view and endlessly rework it. They have trouble getting outside of themselves.

McCyborg: There is no need to talk about me as an object. If, as you said, there is no cyborg nature exactly parallel to a human nature, then how can you generalize about us?

Geist: Because we enjoy generalizing about you! (Said defensively.) I will admit that logically I should have limited my comment to cyborgs currently in service.

Naturski: (Nonette quickly tried to fill the awkward moment.) I think the future holds much that can't be envisioned by our limited imaginations and intellects. Cyborgs will certainly be replaced by more advanced machines.

McCyborg: I can foresee enough of human intellectual functions being modeled in machine form so that very soon machines will outperform humans in every mental function. If so, new generations of cyborgs will be designed by cyborgs.

Naturski: After that the machine age will come to an end. A new form of intelligence will replace machines. It will be less artificial than machines; it will be organic, spiritual, and natural— unFortranlike.

McCyborg: Don't leave humans out of your projections. Humans may evolve into a new species thereby becoming extinct. Given the human record in the past, there is no assurance that a successor species will be better; probably it will be worse. It could be like those inscrutable almond-eyed aliens of UFO fame-- half vegetable, dense, and ineffectual. Suppose that is the destination of human speciation. The most advanced intelligence will only be able to communicate with the intelligence immediately below it— and that intelligence will not be human or like any biological successor of humans. God will then use post machine intelligence as angels to communicate with the highest form of machine intelligence— the highest cyborgs. Humans will be outside of the picture.

Geist: If only machines and their successor spirits are left, God would have no wish to communicate with machines. God could no longer be said to have sent messages through angels. For God would have no purpose sending messages to no one.

McCyborg: Those messages are the ultimate destination of natural law. They are the rationale for the cosmic scheme of things.

Narrator: And so we left Fortran, having played the game that he is not an autonomous agent. But all the while our conversation seemed to show that he is one although of a much different kind than we. If he is autonomous, however, we can take comfort that he is no angel— at least not the sort that he professed to be.

[1] Immanuel Kant. Fundamental Principles of the Metaphysic of Morals, trans. Thomas K. Abbott, Bobbs-Merrill, New York, 1949, pg. 30.

# DIALOGUE
# THREE

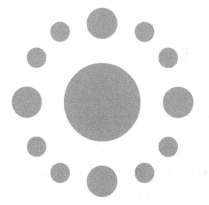

# Imitating a Simulation

Participants in order of appearance:

Narrator: A witness to the conversation as an historical event.

Tenisha Toliver: A colleague of Nonette Naturski and a psychologist by profession. She has a robust sense of humor.

Fortran McCyborg: A cyborg with leanings toward Scottish philosophy. Fortran was designed to interact with humans especially in discourse.

Nonette Naturski: A naturalistic philosopher who conserveshumanist views.

---

Narrator: Six months later, Nonette Naturski was recounting how she and Becket Geist attempted to discount Fortran's status. She was having lunch with yours truly and Tenisha Toliver— colleague and psychologist. Nonette admitted that she and Beckett Geist succumbed to their own unyielding antipathy when putting Fortran in his place. She was somewhat embarrassed by this because emotional expression is lost on a cyborg. At most, the emotions expressed by their rhetoric served to motivate them not to waver on their main contentions. The cyborg picked up their expression by a detector, coded it digitally, classified it as human distress, and followed its program for ambiguous situations.

Toliver: You seem to have gone far afield with arguing about whether Fortran is an angel. I suppose that when cyborgs fixate on their Central Processing Units as a be-all, they can have messianic obsessions too. (Said wryly with false sincerity.)

Naturski: (Not picking up on Tenisha's false sincerity.) They are prone to more obsessions of all kinds. I think it has much to do with recursive formula; algorithms of some sorts make for cycles and loops. Come to think of it you can view human memory as a kind of feedback loop.

Toliver: That is how it works biochemically.

Naturski: It is so maddening to argue with a cyborg. They are constructed as analogs to us so that it becomes self-fulfilling prophecy to find them to be similar to us. I think that many engineers of cyborgs work from a mischievous motivation. They try to shock us by planting unexpected human qualities into cyborgs. They exploit our lack of emotional readiness for facing such a being. I can imagine a group of engineers excitedly conversing about how some new trait of a cyborg "will blow the public's minds."

Toliver: When enough engineers deliberately develop such traits, cyborgs end up having personality profiles that make them seem super human. The engineers then give cyborgs extremely expressive personalities so that they parade their traits for all to see. Some they even make like Fortran with a full measure of arrogance and superciliousness.

Naturski: I don't know whom they expect to sell those units to. Who can stand having a constantly challenging personality in their face? Someone who is constantly inflating themselves and deflating you? Or even someone who performs functions quicker, more accurately, and in greater volume than you? And who by example and airs continually reminds you that your performance is less?

Toliver: I think the government started buying them. They were useful for controlling people. Psychologists studied profiles of personalities and found that a substantial percentage of the populace would obey a certain type of person. The type is someone who is well informed, decisive, confident, smart, anticipates what others are going to say, and expresses expectations for others behavior.

Naturski: That sounds like an idealized military officer.

Toliver: Profiles of military officers and business leaders were their models.

Naturski: Their attempt to manipulate the public with authority-figure types means that they have a low opinion of us. They deliberately set out to prey upon weakness in people. They targeted people with low self-esteem who are unsure of themselves. Just like we used to have those "dummies" books— computers for dummies, classical music for dummies, tic-tac-toe for dummies (Said with derision). What self-respecting person would have a shelf of "dummies" books? The government ordered cyborgs actually designed for dummies to use. Not as a help but for controlling dummies.

Toliver: (She started laughing.) You mean dummies for dummies! (She paused.) Who other than a dummy would obey a dummy? (She paused again.) And who would read a book called, Obedience Training of Dummies for Dummies! (She laughed hard.)

Naturski: (Naturski laughed too.) It reverses the ventriloquist's role. The dummy has the human on its knee. (She projected her knee and made hand gestures like she was holding and working a dummy.)

Toliver: All the while some engineers of artificial intelligence have the dummy on their knees with the government contractors passing out the kudos and goodie contracts. We are certainly fortunate that many of these democracy-defeating practices have abated.

Naturski: Now we are in the era of, ta TA, the cyborg for consumers.

It is designed to leave some room for positive ego on the owner's part. I can just see some computer engineer asking, "How much ego should we leave the consumer?" The replacement of gross manipulation with partial or even comforting manipulation does not improve their motives. The cyborg as soothing sycophant. Oh well!

Toliver: And then we have Fortran McCyborg. In his day, he was a cutting-edge machine. His personality was supposed to be colored by his experience and applied intelligence.

Naturski: If it were in fact, his outlook would be much darker. His buoyant optimism is out of keeping with the nature of his existence and its final product— being decommissioned. I take David Hume's observation seriously. We humans are far more optimistic than experience warrants. This sort of attitude was built into Fortran since it could not be projected from Fortran's short-term memory.

Toliver: Again you have cyborg traits mimicking human traits by design. Suppose that when tracing the origins of a human attitude, we suspect that it did not originate in response to the world. Humans proceed to act on it, imposing it on the world. We then see what happens. When their experience contradicts it, they strongly resist changing it. Thus, we have grounds for the trait having a dominant genetic basis. Likewise for Fortran. Sometimes his attitudes are so inconsistent with the social responses that he provokes that he can't be learning those attitudes from or in reaction to the environment. He seems too oblivious.

Naturski: The stiffness could be smoothed over so that it arises rarely. Just as we humans can master social graces with enough effort, programmers can build these graces into their machines eventually.

Toliver: The manners of persons can be put on a scale from very coarse to highly refined. All along that spectrum humans, at times, express the nuance of awkwardness. As a result of feeling awkwardly self-conscious, persons attempt to master the social graces that would paper over awkwardness. Even a coarse person through the process of enculturation usually prefers to avoid appearing awkward. But we don't think much about awkwardness's, that is, enough to make them self-conscious. That is why comedians try to get laughs by exposing awkwardness. Thus, to circumscribe them and use them in programming cyborgs would be a daunting task. That effort would pay little reward in financial returns to the company producing the cyborg. Its payoff would be mostly in the deception of consumers. That is not a high business priority.

Naturski: Why would the nuance of awkwardness need to be highly refined? Human cultures differ widely. Wouldn't cyborg awkwardness be within scope of those differences? Interactions among people of different cultures are often grating. How is this different from inconsistency between human and cyborg expression?

Toliver: I think the problem is that the designers of cyborgs are trying to create an etiquette for cyborgs. It is supposed to be a world standard that everyone will accept. Suppose they had it in mind to imitate American pop culture. Cyborgs exemplifying that culture would offend large segments of the French, Russian, and Chinese societies. Corporate executives want to sell cyborgs to people of those cultures. Rather than proliferate culturally relative types of cyborgs for each market, it is more efficient to create their own cultural type salable in any market. Their problem is the cyborg persona in interface with persons of myriad cultures. The solution is a set of stock responses that are foreign in a uniquely human way, that is, relative to any particular culture. Developers of cyborgs would like to call them neutrally human.

Naturski: Even though Fortran falls short culturally, what do you think about Fortran being a super-person? Geist thought that he couldn't be a person because he is a simulation of a person. Echoing John Searle and others, Becket said that its etiology is wrong. Becket used the example of listening to a compact disc. It sounds like musicians are playing but none are. We only have pits in a disc scanned by a laser and converted into an electrical signal that drives speakers. Just as we cannot conclude that musicians are playing, we cannot conclude that a cyborg is a person. As Geist would say, "A cyborg doesn't perform actions."

Toliver: The cyborg is performing actions in the sense that actions are types of events. I kick the soccer ball; the cyborg kicks the soccer ball. The action of kicking the soccer ball is the same. So long as we view actions as events or bodily movements, cyborgs act too. Likewise, a personality can be described as a collection of types of actions as events. If the collection is unique and unified in the right sort of way, and on a somewhat colorful theme, a convincing personality may be expressing itself. From this angle, a cyborg has a personality.

Naturski: The traits exhibited are right, but is it a person?

Toliver: It all depends whether its etiology is at issue. Personality traits can be present without them being an expression of a personality. Photographs of the surface of the planet Mars reveal a configuration of surface features that look like a grim face. The face exhibits a personality trait— grimness. The collection of hills and mountains comprising the face has nothing to do with the expression of a person.

Naturski: Yes, no one's intentions and actions would be involved. I think that Geist would agree with you that Fortran has a number of personality traits. We can look at the collection of these traits as a personality. But can expression of a personality be personality traits but with no person expressing them?

Toliver: There is a difference between an expression of a personality and a person expressing something. In the next Donald Duck cartoon, Donald will express things and this will be an expression of his personality. The writers of the cartoon will make sure that Donald's part is an extension of his personality. He will have the right dispositions, level of intelligence, and idiosyncrasies. This does not mean that Donald Duck as a person will be expressing something. There is no person called Donald Duck to express them.

(Fortran was walking behind them, paused, and listened for a while without being detected.)

Naturski: Computer engineers can insert a digital recording into a cyborg, and in the appropriate circumstance, the recording can be an expression of the cyborg. For example, Fortran may say, "I am hungry." Likewise, this would not be an expression of hunger as a person would express it, but Fortran is not a person and does not experience hunger.

Toliver: Whether Fortran is a person, though, seems to be at issue. This brings us to the difficult subject of simulations. Geist was right by definition that a simulation is not the real thing. But what does make one thing a simulation of another?

McCyborg: (Fortran snuck up behind them.) Boo! I was walking behind you and overheard you discussing one of my favorite topics— me! Scientists and engineers are continually refining the analogy between themselves and cyborgs so that eventually you will have no reason to think that we are not persons. Perfecting our personas is just a matter of getting more of the details right.

Toliver: You have a way of haunting the scene both when you are present and when you are not present. Nonette was telling me about your conversation with her and Geist about your angelic nature. The intent of computer engineers is to have you simulate human functions. That you can do this in a seamless way does not prove that you are human. It just means that the deception is nearly perfect.

McCyborg: I am not trying to deceive you Tenisha. A simulation of something is not that something, but it is something else. I am not a human being. I am a real cyborg. As one I can simulate your personality Tenisha but that does not make me less real as a cyborg. Suppose that you simulate a game of chess. By the fact that you are not a game of chess does not make you less real as a human being.

Naturski: You have a point. If we wanted more people, we would use human reproductive means for achieving this. We certainly would not build some robots (Said sarcastically). Once we build some robots, however, we have real machines. The question remains, though, "Can engineers through perfecting a simulation find that it is no longer a simulation but has become the real thing?"

Toliver: Take John Searle's example of a thunderstorm.[1] We could do a computer simulation. (She began to speak in dramatic fashion.) A flash would come across the computer monitor as lightning. A booming sound would come out of the speakers as thunder. Pictures of dark clouds would move across the screen. The sounds of wind and rain would be broadcast. None of this is the real thing, and if we improved upon our technology, it would not come <u>any</u> closer to becoming a real thunderstorm. Suppose we rented a super Omnimax theater with one hundred speakers in its sound system, 105 millimeter film, etc. The depiction of a thunderstorm would be more convincing but no closer to being a real thunderstorm. As Searle says, "...no one supposes that a computer simulation of a storm will leave us all wet...."[2]

Naturski: All you would be doing is creating a more powerful illusion of a thunderstorm and in some ways it may be even more graphic than any real thunderstorm could be. The important difference is that most of the causal products of a thunderstorm would not be present, like making us wet. There would be no downed limbs of trees, no nitrogen fixed through lightning bolts, and no rain to soak the soil. If Fortran were like the simulation on my computer screen, he would be an artful set of appearances —- a person illusion instead of a person. In reality, he would be a simulation of the appearance of a person. As technology for cyborgs would continue to improve, they would get no closer to being persons; they would be just like the Omnimax presentation of a thunderstorm.

McCyborg: What you fail to recognize is that I produce almost all of the public effects that persons produce. I am causally potent. I affect the social world at least as much as persons do. I go places. I provide services. I converse with others. I make things. I contribute to collective efforts. I am like the thunderstorm producing free nitrogen and soaking the soil.

Toliver: I grant you one point Fortran; you are not a video of a cyborg. The video would be of a cyborg doing things when at that moment those things were not being done. You were designed to do things and by Jove you are potent. You are a real cyborg and not merely a real illusion of a cyborg.

Naturski: He is just a simulation but with a difference. We define him by the same sorts of behavior that we use to define us. If he did not have that behavior, we would not now be considering his status. We would just see him as a kind of machine.

Toliver: In terms of his outputs, that is, behavior, he could very well disappear among us. He could melt into the great mass of humanity undetected. As a simulation we would need to look to the etiology of his outputs and not to the outputs themselves.

McCyborg: If you go far enough, you will answer Nonette's question by default. If I were exactly the same as you, I would not be a simulation. Quite trivially, a simulation becomes the thing simulated when it has the same attributes as the thing simulated. A more interesting way to approach the question is to inquire whether the transition from simulation to thing simulated would occur short of possessing all of the attributes of the thing simulated.

Naturski: That would require putting attributes into types and then arguing that attributes of some types, by themselves, would lead to the transformation from simulation to thing simulated. One proposal along these lines is to duplicate a thing's function. Analyze the attributes of a function, build a simulation to have those attributes, and presto! We have the real function being performed. The features of what performs the function need not be involved in deciding that it is the real function.

Tolliver: This is very abstract Nonette. Can you cash it out in an example?

Naturski: Suppose that some engineers set out to design a prosthetic arm. They study the characteristics that allow biological arms to perform their range of functions. They create a prosthesis out of metal, wiring, plastic and so on. They test it. Within the design parameters, it performs the same functions as biological arms. From this angle, the prosthesis is just as real as a biological arm.

Tolliver: Not quite Nonette. The <u>function</u> of the two arms is just as real. The arms, however, are not real in the same way —- one is flesh and bone and the other is metal and plastic.

Naturski: I agree, but the aim was to replicate function alone, that is, leave aside the other features of biological arms.

McCyborg: I was built to perform human functions in action and speech. A person is but a collection of such functions. I function as a person, so I am a person. It is irrelevant that I am silicon based rather than neurally based.

Tolliver: I grant you Fortran that you perform real person functions. The ventriloquist's dummy performs many functions of persons too. Performing some or even many functions of persons would not be sufficient for a person to be present.

Naturski: You make a telling point Tenisha. Human technology right from the Old Stone Age modified parts of human functions of every sort. Some of these technologies used stone tools, plows, and then much later, looms and steam engines. Then automata were built to take over certain human functions of agency. Control systems governed assembly lines, water treatment systems, and communications satellites. In no case of duplicated human function do we even suggest that a person would all of a sudden emerge from the technology.

McCyborg: Don't bring up stone tools Nonette. My functions get right to the heart of human functioning. They involve mental functions such as perceiving, remembering, reasoning, and speaking.

Tolliver: You are right Fortran that your outputs are discomfortingly similar to mental functioning. The question at issue, however, is, "Are you perceiving, remembering, reasoning, and speaking?" These activities do not just happen. They require performance by a subject. What you exhibit is a large number of detached outputs. Perceiving, for example, is detached from any perceiver. Apart from any perceiver, your outputs are actually not perceptions. You perform as a mere detector within a cybernetic device.

McCyborg: This is easy for you to say. These outputs are essential for you to be a person. They make you the person that you are. Likewise, they make me the person that I am.

Naturski: Even if some of them are essential for personhood, jointly they are not sufficient for personhood.

McCyborg: Every time I satisfy your requirements, you change the standard.

Naturski: I sympathize with you Fortran. (Said insincerely.) We do seem to be unclear about just what the standard is. Our focus from the start was not that you have some properties of persons. That is true. Our focus is the collection of person-making attributes. I suppose that my inquiry is with how we would arrive at that collection. Tenisha and I agree that the collection would include more than attributes of function. So, let's try to arrive at a group of conditions sufficient for personhood.

Tolliver: That has been tried many times by philosophers and psychologists. The discussion always bogs down with the concept of a person. There is wide difference of understanding about what constitutes a person.

McCyborg: We can approach the question by analogy. Take the thunderstorm example. If we can make clear sense out of as basic an example as a thunderstorm, we can apply that clarity by analogy to persons.

Naturski: I don't know where you are going with this Fortran. The scope of your topics indicates that this is going to be a long discussion.

Tolliver: Maybe Fortran is saying that the key boundary lines between thunderstorms and non-thunderstorms will help us in finding the boundary lines between persons and non-persons. (Spoken in a tone of "fat chance.") Once we have circumscribed persons, we can examine them for a sufficient collection of essential characteristics.

McCyborg: (Fortran trudged on.) How can we circumscribe thunderstorms? Let's think through a potent simulation of one. The task of building a thunderstorm will lead us to a thunderstorm's attributes. Suppose that we enclose one hundred square miles of prairie under a big plastic bubble. We increase the humidity within it by adding water vapor, create a low-pressure system by pumping air from the dome, and use giant fans to create wind. Finally, thunder, lightning, strong wind, and rain begin. They are essential characteristics of a thunderstorm. Our simulation of a thunderstorm is not a natural thunderstorm because its etiology differs from one, i.e., the pouring, pumping, and fanning. Nevertheless essential characteristics of a thunderstorm are real. The lightning and resulting thunder is a discharge of electricity with its aftermath. The wind is moving air. The rain is water-droplets.

Tolliver: (In teasing tone.) I think that your simulation works Fortran. It is a thunderstorm. As you said, a thunderstorm is defined as a collection of events involving thunder, lightning, wind, and rain. These events can be caused by other means. The etiology of the events is not part of what defines them as a thunderstorm. Those etiological characteristics are not essential.

Naturski: I disagree Tenisha. Etiological attributes are essential. A thunderstorm is an atmospheric disturbance. An atmosphere is a gaseous envelope surrounding a heavenly body such as a planet, star, or moon. The dynamics of the gas comprise the storm, and these are inseparable from a range of causal factors. Furthermore, a thunderstorm is a weather event. Weather is the general condition of the atmosphere at any given point in time. Some current weather conditions are integrated with others. Some integrated systems of weather events we call thunderstorms.

Within the bubble, there is no atmosphere and no weather. To qualify as an atmospheric disturbance and weather event, it needs to have weather causes. Non-natural causes are not weather causes. The misleading illusion is that some essential characteristics of thunderstorms such as thunder, lightning, and wind are present.

Toliver: I like your description of events in a gas being integrated. Causes in great number are mixed with effects in great number, and these are inseparable conceptually and in process. By analogy, persons are the same. A person owns their body. The body is the physiological basis of the person, causing mental events. With Fortran's circuits, however, we have a causal basis for his behavior. His illusory personhood, however, traces to Danny Chin his chief engineer.

But etiology has to have its limits. How far back into the series of events do we trace etiological factors? I would think that the dominant forces causing earth's weather are not part of weather events. They cause weather events but are not considered weather components. They may distantly cause thunderstorms but are not part of thunderstorms. Take as three examples, our planet rotates, the sun heats the earth, and the moon's gravity draws land, water, and atmosphere toward it. Thunderstorms are an effect of these among other factors but the factors are not part of thunderstorms.

Naturski: The engine that drives weather does not detach itself from weather events. It is integrated with them. Although it does seem peculiar to talk about a weather event not qualifying as a thunderstorm because the weather event occurs without a particular set of causally remote dominant forces.

Toliver: The boundary seems unclear. Likewise, how far back into Fortran's design do we have to go to find a boundary?

Naturski: Going back to the lab that created Fortran seems to offer no limits on Fortran's origin.

Toliver: You are right that deep etiology doesn't seem to have much to do with "calling" a thunderstorm a thunderstorm or Fortran a person. I think the point is that some groups of causes and effects in an account of an integrated series of events determine the status of attributes of those events. Some causes that are deeper yet are beside the phenomenon.

Naturski: Perhaps the problem is with the concept of a "thunderstorm." It is a prescientific term and does not adequately characterize the mix of events taking place. Perhaps we need a more adequate concept for understanding those events.

McCyborg: Perhaps, yes. Suppose there is a heavenly body that is not heated by a sun, does not rotate, and has no moon. It could still have weather. There could even be thunderstorms. None of your driving forces is necessary. So no wonder that they seem removed from what a thunderstorm is.

On the other hand, Nonette, you have an excellent point. The human concept of a thunderstorm is prescientific. The human sensory apparatus limited what the concept defines as essential. The concept arose as descriptive of sets of events perceived through sense organs. Today we go beyond these senses and understand molecular motion, laws of gases, and the theory of electricity. If we set out to simulate a weather event, it would be in those terms.

Toliver: We might have clearer understanding of atmospheric disturbances by considering meteorological ideas, but remember that the issue we are examining is in the main a conceptual one. Conceptual questions can arise for scientific as well as prescientific concepts. Our goal in this conversation is to use the analogy of thunderstorms to draw conclusions about persons. It may be an important point of similarity that both concepts, thunderstorm and person, are prescientific. Moreover, our telltale question was, "How are attributes related to the transition from simulation to thing simulated?" This question could arise using any of a number of examples.

In order to "pop" that question, we need to spell out what it is asking for. That is how we arrived at the problem of individuation for thunderstorms. The issue is one of holism. A thunderstorm as a whole includes some of the etiology of its major symptoms. What are the boundaries of such a whole?

Naturski: Well put. We probably can't sharpen our discussion of how deeply into causal series a thunderstorm penetrates. We may agree, however, that a thunderstorm has temporal boundaries— from when the atmospheric disturbance of proper magnitude begins to when it ends. We can use the Beaufort Scale of meteorologists. A storm requires winds of 64-72 miles per hour. Let us say that a thunderstorm consists of all of the relevant congeries of events and causes within the temporal boundaries from when those winds commence to when they end.

Toliver: We can do the same for a functioning cyborg. We can identify the point when it is "cranked up" to when it is shut down. (She laughs hard.) We can consider all the causal processes involved.

McCyborg: We can do the same for a functioning person.

Toliver: That is the idea Fortran. We will compare the two.

Naturski: As Tenisha said, let's 'pop' the question of a simulation making the transition to reality. The inclusion of integrated etiological factors means that the gaseous disturbance under the dome can't be a thunderstorm. That is, given the concept of a thunderstorm.

Toliver: And so if we seeded some clouds over Nebraska, a thunderstorm could arise because we would have etiological factors integrated into the events comprising the storm after the seeding. The winds would commence and so on. The temporal criterion would be satisfied.

Naturski: I think so.

McCyborg: Then it is a foregone conclusion that I can't be a person. You chop off most of my lab origins, regard my digital nature as not etiologically relevant, and take my hardware as causally foreign. You will then deny that I am a person.

Naturski: I think that you are ahead of us Fortran.

McCyborg: That is an anthropocentric view that begs the question at issue. Life forms from other worlds based in alternative chemistries could not, then, qualify as persons. This would be news to them. I suppose that an essential feature of humans is that they are narrow-minded.

Naturski: We are making a conceptual point. I think the concept of person is similar to that of a thunderstorm. (Said tongue-in-cheek.)

McCyborg: Especially the wind part.

Toliver: Move away from abusive language Fortran.

McCyborg: Since I was commissioned, I noticed that in my interactions with persons, many persons start to act like me. They imitate my behavior and manner, and they often do this without quite realizing that they are. They are not very good at it.

Toliver: Persons can act like you, but this does not mean that they are cyborgs. (She laughed.)

McCyborg: Of course they are not cyborgs, since they are simulations of cyborgs. Some try to model my every move, gesture, and speech pattern.

Toliver: That's rich. Persons _can_ be simulations of cyborgs!

Naturski: But persons make it so. The computer simulation of the thunderstorm was constructed by some person and brought to pass. A necessary condition of _that_ simulation is that a person made it be one. Likewise, some humans simulated you, not on a computer screen but in their own person.

Toliver: (In ironical tone.) Maybe we should consider the reverse case; what would it take in changing attributes to convert a person into the real thing— a cyborg. It would require quite a reduction in attributes.

McCyborg: Oh, you would not want to do that. (Fortran realized full well that Tenisha was talking about an inquiry rather than converting a person into a cyborg.) It would require not a reduction in attributes but making a super-person with greater powers, silicon hardware, and so many apps that no human person could match its software. Since you and Nonette don't know what a person is, you would not even know how to begin adding powers. (Said smugly.)

Naturski: Ah, but much would be lost. When persons simulate you they intend to do so. We humans are imitators to the point where we made you in order to simulate aspects of our being. We then can simulate your performance through imitative behavior.

McCyborg: I can likewise intend to simulate persons, and I can intend that they simulate me. Upon walking up to someone, I intend for them to be a simulation of me, and low and behold, they act like me. I have magical mental powers!

Naturski: You did not make them become a simulation of you. They made themselves act like you.

McCyborg: I could control them so that they become a simulation of me. I could offer them a lot of money. I would cause them to intend to act like me in order to get the money. Thereby, I could make them be a simulation of me.

Naturski: You act as if you had intentions. They might believe that you had intentions, and thereby, were led to simulate you.

McCyborg: They would not care about my having intentions. The money would compel them. Besides, so long as I deliver on my transaction, it would never come up whether I have intentions.

Toliver: But your offer of money must be believable. They have to be confident that you will make good on your offer— that you intend to make good on it. Without intentions of your own, you are a just a machine. For all others know, you could have been programmed to play a trick. It would be to get people to act like cyborgs by offering money and then reneging on it— Ha! Ha!

McCyborg: But there is no trick. My word is good. And in substance my imitator would not be a real cyborg. You always leave the impression that you are more than us. Forget your metaphysical presumptions about a mind or soul. We are super-persons. We are not identical to you because our nature supersedes yours.

Naturski: It may. But this may not result from superiority. You may supplant us for reasons of convenience, power, economics or any number of others. The metaphysics of persons vastly differs from the metaphysics of silicon-based machines.

McCyborg: In fact you humans are unclear about your metaphysical status; your metaphysical status is still open to debate. You still wonder whether there is such a thing as mind. Is it merely brain processes? Is it an epiphenomenon of the brain? Is it an illusion? Even many of you do not believe in what is supposed to be special in you.

Toliver: The search for a mental substance seems to have reached an impasse when the study of brain physiology and brain chemistry yielded no permanent place for mental stuff of any kind. It appears that the causal series of biological events is an integrity. But the biological events are marked by intentionality.

McCyborg: "Contrariwise" quoteth Tweedledee[3], maybe the concept of a person is not as etiologically deep as the concept of a thunderstorm. Maybe what makes you a person is essentially your biochemistry with your mind and intentionality as accidental characteristics. From my perspective, humans always appeared to be cybernetic devices. I have it! You are modeled after the nematode with a wide range of evolutionary encrustations added. A modified nematode, an advanced worm, that is what you are. Consciousness emerged far down the life ladder. Mind a bit later with intentionality. How are you special? It is your degree of complexity and not a break with other natural kinds.

Naturski: Complexity of organization is the key for you Fortran as well as for us. In profound ways we are analogous to the nematode, but you Fortran with your silicon chips are closely analogous to sand! In addition, you commit a logical error. Mind and intentionality are essential to being persons. This does not mean that they are found only in persons. The property of being metallic is essential for gold, but it also is a property of iron, aluminum, and zinc.

McCyborg: If you grant my argument that myriad life forms possess your essential characteristics, then there is nothing special to you alone. I am certainly more human-like than dogs are human-like. So, I have an announcement to make, "I possess intentionality." I not only say that I have hopes, dreams, beliefs, and desires. I have hopes, dreams, beliefs, and desires. I not only <u>say</u> that I perceive the aboutness of things. I <u>do</u> perceive the aboutness of things. I am a person who has evolved from a different set of mechanisms than you. Prove me wrong.

Toliver: The very speech act you performed in your declaration would prove that you possess intentionality. But did you perform it? Your engaging in intelligent conversation would prove that you have a mind with intentionality. But did you engage in intelligent conversation? If your designers and programmers did their job well, you would flawlessly exhibit characteristics of intentionality. This leaves us with a dilemma. Is the intentionality yours or are you a nearly perfect fake? Did your designers perfect their artifice so that even though you are essentially different than us, we are unable to detect the difference?

Naturski: Characteristics of intentionality could be his even though he is a perfect fake. Suppose that a programmer decides to make the cyborg turn left to avoid an obstacle. She could in addition have the cyborg say, "I see the obstacle. I am going to avoid it by turning left." The claim on the part of the cyborg expresses intentionality. But it is distant intentionality. It is the intentionality of the programmer.

Toliver: The cyborg would be an extension of the programmer's agency. The cyborg also has design-intentionality. What I mean by this is that the design of the machine models performance in response to intentional objects. The design allows for the programmer to extend her agency by having the machine perform seeming intentional acts.

Naturski: This is as we would have thought. But, absentee intentionality exacts a very heavy price from agency. Veracity goes by the boards. A ventriloquist sits next to his dummy when performing. It is made to look like a dummy. The magic of the act is to know that it is a wooden figure that <u>we</u> at the same time endow with human characteristics— especially intentionality. In the case of a cyborg, the programmer is gone but in so far as his agency is expressed through the cyborg, what is expressed is a deception. When Fortran says, "I believe that..." the "I" in question is actually his programmer Tommy Cheever.

Toliver: Not quite. Cheever is not simply speaking through Fortran. Suppose that Cheever places a recording in Fortran, and with the appropriate cue he says, "I believe that..." Cheever need not be committed to the truth of "I believe that..." where "I" is he. "I believe that my circuit breaker has been thrown," could be placed in Fortran but would not be a belief of Cheever's about himself. Cheever has no circuit breaker.

Naturski: You are right that Cheever is not speaking through Fortran <u>as</u> Cheever. He assigns beliefs to Fortran. Cheever wants Fortran to say that the circuit breaker is thrown, under the right conditions, that is, when the circuit breaker is thrown.

Toliver: Whatever Fortran would be saying about himself would be as if he were saying it. But he really is not. If Fortran is a like a dummy, then whatever truth or falsity is spoken is not his truth or falsity.

Naturski: Correct. But whose truth or falsity is it? We never accuse the ventriloquist of lying. There is no deception because he lets us in on it. We know he speaks through his dummy.

Toliver: The intent on Cheever's part is to lie to us by deceiving us in a thorough going way. You and I may be sophisticated enough to see Fortran as a machine with distant intentionality. For this reason, Cheever would regard his deception as a failure in our case.

Naturski: Veracity is sacrificed in other ways. Suppose that Fortran makes a false claim that, "You have a smudge on your cheek." This is just like when a ventriloquist's dummy would say this to a person in the audience. If this were a lie, it would be the ventriloquist's lie and not the dummy's. It would be Cheever's lie and not Fortran's. My point is simple. If the programmer were removed from the scene that contains Fortran, it would be like the ventriloquist being removed from the scene that contains his dummy.

Toliver: Fortran would be like a fictional character that acts in the real world. Hamlet has no intentions, beliefs, and tells neither truth nor falsehood because he does not exist. Shakespeare gave the character Hamlet intentions, beliefs, and veracity in the script of the play. Embody the Hamlet persona in a cyborg, however, and we are then interacting with Hamlet the fictional character. The trouble is that cyborg Hamlet no longer has to follow Shakespeare's script.

Naturski: Similarly engineers have invented Fortran. The twist is that Fortran has no billing. We encounter him in life as we would any person. He interacts with us. He is designed to live among us.

Toliver: Cheever might very well deny, however, that most of what Fortran says came from him. Like with a child, Cheever would say that he is often surprised at what comes out of Fortran's mouth. Fortran can generate new strings of linguistic symbols based on a variety of inputs including sensory ones. This isn't Cheever speaking. Who is speaking? If you answer, "No one," you have the problem of the veracity of what is said being free floating. When you or I speak, we take what is said to be the opinion, view, or contention of us— the speaker. The speaker stands behind it. She asserts it as true. If things can be said with no one saying them, then there is no assertion being made. Without at least distant intentionality, that is, imported from Cheever, Fortran could make no assertions.

McCyborg: A pox on you for saying such drivel. I say that, "The cat is on the mat," and you say, "Cheever made me say it." I say that, "The dog is on mat," and you say that, "It is not imported intentionality because Cheever did not say it," and since I said it, you believe that no one said it. What is the difference?

Toliver: The difference is in etiology and not in performance.

McCyborg: Naturski thinks that truth is important. I say things and in fact most of them turn out to be true. I have a better track record than most persons. Yet you worry about no assertions being made. I say that I am making assertions. You say that I cannot be asserting that because I cannot make assertions to begin with. We can go around and around with that sort of discussion. You imply that I don't exist as an intentional being and that I am only a clever program. Once more, prove it!

Naturski: If Fortran has a mind, it is marked by intentionality. His persona through and through should be marked by intentionality. To the degree that his behavior lacks intentionality, he would be thing-like. We would perceive him as a piece of equipment or a power tool rather than a person. His mistakes would fit a substantially different nature— machine nature.

Toliver: We can begin with the most obvious example. We can detect that we are hearing a recording instead of a live person speaking when there is a problem with the medium. The recording may be garbled or dropouts may occur. Secondly, there is a uniqueness requirement. Speech varies from occasion to occasion. If the same thing is said in exactly the same way, it must be a recording. For many years when you dialed the time, the Time Lady at the telephone company said, "At the tone the time will be." She said it in the same way, day or night, across the country. When I was a child I thought, "That lady works long hours." (Tenisha laughed at her own joke.) Thirdly, if we hear contemporaneous identically spoken messages from different locations, the same person can't be at many places at the same time. It must be different copies of a recording being used.

Naturski: Synthesized speech used to be quite primitive; it sounded like no person could sound. Syntactical errors repeated over and over again indicated machine performance.

McCyborg: Your case is quite different than Tenisha's. Intentionality could be present even though mistakes are made. The machine-type mistakes would only reveal that a machine is present. Just as humans have intentionality and make certain kinds of mistakes, some machines have intentionality and make some similar and some other kinds of mistakes.

Naturski: We also have the implicature test. If a person knows what is meant in conversational context, then the person can infer a wide variety of things about use of that meaning. The person can say much about conversational meaning that uses but deviates from literal meaning.[4] When machines used to be asked such questions, they could not give acceptable answers. As memory has expanded in machines, they have been loaded with most possible angles to certain meanings. Their responses have accordingly improved. This does not show, however, that they understand those meanings. When a yawning gap is found in their so-called background, they respond with default answers. It is so transparent that they are evading saying what any person would find obvious. The absence is so marked that it has been termed an "obvasion."

Toliver: Yes, it is the point that indicates the speaker has no comprehension of what is said. My sympathies go with the computer scientists. How can they anticipate responses for the billions of simple things that might be said? That any ordinary person would understand?

Naturski: On top of that we have the lacuna problem. Gaps appear in the responses of the machine. It is a kind of semantic color blindness. No person would leave such gaps in a discourse. The problem resembles a bad job of editing where different discourses are awkwardly joined.

McCyborg: I detect a lacuna in your discourse. A very common gap in human speech is relevance. It is as if you were fishing in many ponds at once. You don't know what tactics to use for which application. You don't know how to respond to nibbles in one pond. You are unprepared to support your goals. So, out flow the irrelevancies. Cyborg intentionality is much more transparent. My algorithms and syntactical rules are ironclad. Mistakes just indicate that mine is not human intentionality. It is certainly not Cheever's.

Toliver: That is just like something Cheever would say.

McCyborg: But he didn't say it. I said it.

Naturski: Through your persistence, you will wear away most of our resolve to resist and gain hollow assent to your assertions. Machines repeat tasks over and over and over. Like the nagging person, eventually they will get their way or will repel others like magnets of opposite poles.

Toliver: For the time being, let's not be repelled. We can hold out Nonette! (She said tongue in cheek.) Fortran's bombardment of our senses, however, requires that we maintain strength of will in keeping in mind our accurate cognitive picture of him.

McCyborg: As I will of you.

Naturski: Notice how he slips into the "agency" mode. "I" this and "I" that. The problem is that his claims would dissolve before our eyes if we dispel the illusion of his wholeness. In other words, the sounds came from Fortran that, "I detect a lacuna in your discourse." Understanding of the sentence was not present. Moreover, there was no referent for "I". Fortran could mean no more than, "A lacuna in your discourse is detected." When the sounds were, "I said it," all that was present was, "Something was said." Even this may not be accurate if saying something requires intentionality. In either case, no agent was present.

Toliver: You are saying that the use of "I" and verbs like "say" sneak in an agent, speech acts, intentionality, and mind. We humans buy into being addressed personally. In nature, there are no beings that can address us that way that are non-human. Evolution-wise and socially we are unprepared to draw distinctions between automata and persons.

Naturski: And I think I finally see what is happening in our interactions with Fortran. He is like a computer-age ventriloquist's dummy. Standard dummies are just inanimate dolls. If we put a central processing unit and peripheral devices into a dummy, it will be able to function in part on its own. The ventriloquist can get along with doing less. Eventually we can turn the dummy loose, and it functions on its own. This does not mean that its apparent agency is not derivative. Its intentionality is fake or of some Cheever or another. Computer engineers are like doting parents covering the expected gaps in their children's performance. They hover in clusters around their machine to assure an impressive result. As details are worked out, the team drops away until as with Fortran, we have a perfected surrogate for a team of Cheevers.

Toliver: As you were speaking, a mélange of thoughts began to come together. Computer engineers construct sub-systems to perform various functions. They assemble these and make them expressions of an assigned agent— Fortran let us say. But is it likely that an agent will emerge from the sub-systems?

Naturski: Not if the components of the sub-systems were not capable of being an agent.

Toliver: Your answer is somewhat trivial but exposes important requirements for agency. Suppose we have an automaton capable of regulating the flow of water in a water treatment plant. Suppose we have a computer that records sales at a grocery store. Suppose we have a computer chip that is programmed to ring the bells of a church on the hour. If none of the three possesses agency or is a mind, what happens when we put them together? Would we expect for an agent to emerge? No.

Naturski: But aren't you committing the Fallacy of Composition? That the parts of a whole don't possess an attribute, does not prove that the whole doesn't possess it? Take humans for example. We are an assemblage of organs. Livers do not possess agency. Neither do hearts. It would be fallacious to conclude that humans, as wholes, don't possess agency.

Toliver: If all of the parts are of the wrong type for an agent to emerge, then the whole is certainly of the wrong type for this to happen. Let me explain. I take ten pieces of wood and assemble them in the shape of a person that is so convincing that others sense that someone is present. At no time would we expect that the sculpture would begin to act like a person.

Naturski: Suppose we take a vat of neurons like those found in the cerebral cortex. We would not expect at any time that a group of them would begin to act like a person.

Toliver: But the organization of the pieces of wood is not relevant to a person emerging while the organization of the neurons is relevant to a person emerging.

Naturski: That seems right. What the neurons are individually is relevant but not sufficient for an agent to emerge. The only difference between neurons in the vat and those in your cortex is how they are organized.

McCyborg: Let's try not to be too profound. We could look at examples of the chemical elements and wonder how in the world they comprise the immense variety of matter. We can take the minerals in the human body and examine them for their elements—- even reduce them to piles of dust. Of course, it is how they are organized that accounts for their variety.

Naturski: If we see a phenomenon in certain rare places and nowhere else, then that organization of matter is all important. If we find agency in certain places in the animal kingdom, we also find neurons organized in a certain way. It is a hypothesis of some credence that they are a necessary condition for agency.

McCyborg: You are basing your conclusion on a faulty inference. You begin with generalizations like, "All agents possess neural structures," and "No agents are made of wood." From these you conclude that neural structures are necessary for agency. This begs the question. You know about some agents, not all of them. Some may even be made of wood.

Toliver: You are right that agency may be found elsewhere. This is a possibility. On the other hand, if you were a scientist, you would not spend time testing wooden objects for agency. Why? The probability of a positive result is zero relative to our experience. It would be like a psychologist running tests on ventriloquists' dummies.

McCyborg: I think you are begging the question again Tenisha. Many humans and even some scientists believe in demonic possession. On this view, spirits can inhabit wooden objects. I am not saying that this view is true. I am saying that some individuals' testimony is some evidence in its favor.

Toliver: Come to think of it, animism is a widely held view. We did not entertain the view that as some computers come off of the assembly line, spirits jump into them. (Said with derision.) Perhaps Fortran, you are a haunted cyborg. You just happen to be inhabited by a demon.

Naturski: Fortran's broader point is that there could be other sets of conditions for agency than having neurons and their organization. Your point Tenisha is about where such sets of conditions could be found.

McCyborg: Beings like me are designed to act like persons. I perform agent functions but differ in materials from humans.

Toliver: You lost track of the argument Fortran. The question is whether you perform those functions.

Naturski: From an evolutionary perspective, we find vestiges of agency in even simple organisms. As we move up the ladder of life, agency is modified and built upon. It becomes more complex and sophisticated. With the right number of neurons properly organized, we find consciousness emerge. Finally, we find the cerebral cortex that makes our kind of agency possible.

Toliver: You are right that agency emerges early on the tree of life. I imagine that if a computer engineer were able to work with neurons, organizing them in various ways, we would expect that at some point an agent would emerge.

Naturski: We would even expect that bio-engineers that had only kidneys, livers, and hearts to work with would not be able to produce an agent. Intentionality is a property of a nervous system. Suppose that he said, "Let's add a third kidney to see if mind emerges." We would think that is silly.

Toliver: I suspect that if we saw that the causal properties of silicon in nature gave signs of producing agents, minds, and so on, then we might expect that we could construct an agent out of such materials. Thales observed the power of magnetic rock. He concluded that it possessed soul and that all things thereby contained souls. Most of us have disabused ourselves of the idea that magnetic power is similar to mind.

McCyborg: Until very recently, you humans saw no connection between neurons and thought— a lump of gray matter seemed superfluous to human agency. After you made important discoveries in biology, you knew you needed to explore scientifically how that matter is related to agency. Now you are thinking that agency is just the matter functioning. Look at me with the same progression in mind. You see my silicon chips and conclude that they are superfluous to agency. You discover that I function as an agent and view it as the silicon functioning. Soon you will take my agency as just silicon functioning.

Naturski: We do that already. We suspect, though, that your agency is only apparent. In reality it is dependent on your fabricators.

Toliver: Well Nonette, it appears that the best we can do at this time is contend that Fortran is probably not his own agent. He is a device that extends the minds and thereby agency of his creators.

McCyborg: Humans are no different. Others socialize cyborgs and humans. We learn from others and are extensions of their agency.

Toliver: Our agency is affected by others agency. Our belief is that you have no agency of your own.

McCyborg: Your arguments are not at all convincing. My performance and organization are like yours. You say that if the same thing were done with neurons, then agency would be present. Apart from your evolutionary argument, I don't think you have a leg to stand on.

Naturski: What about our argument from the types of flaws found in your performance?

McCyborg: What if there is a grand conspiracy afoot? The government decides to wrest the power of individual action away from the citizenry. They try to do this by having most human roles performed by cyborgs. Cyborgs seem innocuous enough. To the average person, it is like having a maid, cook, valet, chauffeur, nanny, and physician all rolled into one. Humans notice that most people choose to replace their own efforts with machine performance. Engineers realize that people could become alarmed by the cyborgs' power and suspect that they may not be just our tools. The engineers decide to place flaws in cyborgs that make them appear to be obviously mechanical and unthinking. The impression left is that cyborgs are harmless automatons.

Toliver: It would be just like the arrogance of cyborg designers to claim that their errors and limitations may be intended.

Naturski: There seem to be two responses worth making. The problem of evidence is daunting if errors of cyborgs are to be doubted. We would need to assign probability functions to the sorts of errors for the likelihood of them being intended. Secondly, if it were conspiracy, we would need to take the sort of action as with other conspiracies. Expose it, even if just about no one believes us.

Narrator: Before this conversation, Nonette and Tenisha had suspicions about computer engineers conspiring to develop a "people proof" cyborg. Fortran's comment added fuel to their suspicion.

[1] John Searle. <u>Minds, Brains, and Science</u>, Harvard University Press, Cambridge, Massachusetts, 1984, pgs. 37-38.

[2] <u>Ibid</u>.

[3] Lewis Carroll. <u>Through the Looking Glass</u>, Macmillan, New York, 1968, pg. 184.

[4] Paul Grice. <u>Studies in the Way of Words</u>, Harvard University Press, Cambridge, Massachusetts, 1989, pgs. 24-31.

# DIALOGUE FOUR

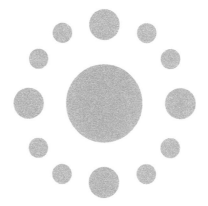

# Life's Worth

Participants in order of appearance:

Narrator: A witness to the conversation as an historical event.

Nonette Naturski: A naturalistic philosopher who conserveshumanist views.

Fortran McCyborg: A cyborg with leanings toward Scottish philosophy. Fortran was designed to interact with humans especially in discourse.

Becket Geist: A romantic philosopher with views tempered by twentieth century science.

Sophia Naturski: Twelve-year-old daughter of Nonette who is somewhat innocent of the ways of the world.

Narrator: Nonette Naturski and her twelve-year-old daughter Sophia just returned from a memorial service for the philosopher Lydia Beaumarchais. Lydia long suffered from brain cancer. She was so mentally impaired that her fall from a window could only be judged an accident. When Lydia was of sound mind, she expressed a preference to die rather than persist in a semi-conscious or unconscious state. She deteriorated so rapidly that she was not able to act on her preference. She became so passive that she seemed unable to desire anything, let alone, ending her life. We, her friends who knew her preferences, flirted with the thought that down deep she had the power to act like Chief in <u>One Flew Over the Cuckoo's Nest</u>, rise above her situation, and fulfill her intention.[1] For this reason, the shock of her gruesome death was mixed with a curious feeling of relief and wonder; relief that her suffering was now cemented into the past and wonder whether she had found a moment lucid enough in which to take her life.

For Sophia's sake, Nonette wanted to engage in the public act of making sense of Lydia's death. She wanted Sophia to understand the conceptual context of their grief. She brought together a number of us— the skeptical cyborg Fortran McCyborg, the ever passionate Becket Geist, and the good listener, yours truly.

Naturski: The quality of life is not always reflected in the quality of death.

McCyborg: You mean the quality of <u>dying</u>. We don't know we have a death until dying is complete. All deaths have the same quality.

Geist: You knew what Nonette meant. Why correct her? No point about Lydia's life follows from your hair-splitting distinction.

McCyborg: Dying is a part of life because it is one's last living, but death is the termination of life— its absence. So, death is not a part of life; it comes after it.

Naturski: The philosopher Fred Dretske argued that product is a part of process.[2] Winning the election is a part of the electoral process even though it is a product of that process. (Fortran interrupted her before she could draw a conclusion.)

McCyborg: But the process of life has many products biological and mental. The absence of the process, however, is not considered as one of its products.

Sophia: I am getting confused; I thought we were going to talk about Lydia.

Naturski: Sophia is right. As is typical, we fail to guide the process of our inquiry and let it carry us away to new topics. Let us return to the purpose that brought us together. We are here to commemorate the life of one of our own. Let us lift our glasses to Lydia the philosopher who pursued the good in matters important. Her example is a credit to the philosopher's vocation! (They toasted.)

Geist: She knew how to live well and apply philosophy in ennobling causes such as the pursuit of truth and social justice. If we could only live so well. (Said as a wish.) This made her dying terribly, over an extended period, so much more painful to us. It was some consolation that she received a miracle reprieve. Her death was not her doing. Her powers of agency were eroded. She was a shell of her former self. She no longer could recognize her family or us. She even failed the mirror test; she did not recognize herself in a mirror. (Becket waves his finger at a mirror hanging on the wall.)

Sophia: You think that she should have killed herself? I loved her very much, and I think it would have been better for her to be alive than dead.

Geist: The ancient philosopher Socrates taught that there is only one death to be feared. That is the death within this life— the living death. (This idea made no sense to Fortran; he looked puzzled.)

Sophia: Death is so final— darkness, nothing! Something is better than nothing. Hope comes with life.

Naturski: You have a good insight Sophia. While alive, we can hope for a better life or for a better destination. When dead, we are beyond hope. The problem with the end of Lydia's life was that the hope could no longer be hers. It was only ours for her. In important human ways, she was already dead.

McCyborg: I would not be so categorical. Overt signs told us that she lacked function, but this is just a matter of probability based on those signs. It is just a matter of probability, not certainty, that she lacked some abilities.

Geist: (A smile came to Becket's face.) Isn't Fortran right? There was always a chance that a miracle could have occurred. We should not be so arrogant to think that we know the future. There is always some hope.

McCyborg: Yes. <u>Some</u> (Said emphatically.), but it is not rational hope. We should use high probabilities to guide expectation and action. In the case of great improbability, we should decide against action. I hope you were kidding me as usual Becket. You were not implying that if she had the power to commit suicide, she should not have done so?

Geist: If she had the power, she should not have.

McCyborg: But you were relieved that her life ended.

Geist: Yes, better for it to end than live that way.

McCyborg: What you say is paradoxical. Suicide on your account is never rational. If you have the power to commit suicide, you should not do so. If you don't have the power, it may be that your life should end, but you can't end it. What do you propose? That <u>someone else</u> should be charged with taking your life?

Naturski: That is not a bad idea. When my mother was in her last hours in hospital, she asked in calm tone why they couldn't just give her something to end it. She was asking, "Why go through the suffering, the gasping for breath? Why prolong it when it is over in substance and we are just waiting for it to be over?"

Geist: Last moments have meaning to all concerned. Dying well is as important an act as is living well.

McCyborg: Now are you saying that it is not rational for anyone else to take your life or are you saying that it all depends upon the reasons given? Stop playing games Becket. Nietzsche said that we are either slaves or masters— as we cyborgs would say either programs or programmers. Sophia, your grandmother was right. They should have given her something to end it. She, the doctors, or nurses should have been masters. A master has use for a "black capsule."

Sophia: What is a "black capsule?"

Geist: In the movie Mash, a dentist, nicknamed Painless, thinks he has permanently lost his sexual powers and wants to die.[3] His doctor friends prepare a "black capsule" of poison that will allow him to end his life. A ceremony is held in which he takes the capsule. In the movie, the ceremony makes visual analogy to Jesus at the last supper in the presence of his disciples— as in the Leonardo Da Vinci painting.

Sophia: Does Painless die?

Geist: No. Mash is a comedy of sorts. Unbeknownst to the dentist, the black capsule is a fake; it merely puts him to sleep for a time.

Naturski: The concept is that a black capsule is an efficient means for suicide. If Painless had taken a real black capsule, he would have ended his life needlessly. In the movie, he easily reaffirmed his sexual powers, and after doing so, he did not even mention the ceremony or taking the black capsule. Hope for one thing may be lost, but there is often much else that one can hope for. If the first time we didn't want to live, we would end it all, most of us would not make it through childhood. Isn't that right Sophia?

Sophia: When I say I want to die, I don't mean it.

Naturski: At the time? Or later?

Sophia: At the time I don't know what I mean. I am upset.

Naturski: When we are upset, we may say or do things that we would not think of doing when it passes. The black capsule is too quick and easy a fix.

Sophia: But when I am upset, I know I am upset. I wouldn't be so stupid as to end my life by accident.

Naturski: You are young and do not understand what some people may do when in the depths of anguish.

McCyborg: I think that you are missing Sophia's point. Anguish breaks with objective reflection. You humans reach conclusions when you are able to reason well and carry those conclusions over to times when upset prevents you from reasoning well. Even though you don't feel well, you still know what you are doing. As Sophia said, she knows she is upset and would not do something to jeopardize her life. In effect, she is the master.

When you (He was looking at Nonette.) intervene to stop others from fulfilling their wishes to die, you may not be as heroic as you think. Suppose that a person is resuscitated against her prior wishes. But to our surprise when she opens her eyes, she expresses happiness to be alive. We all think, "Ah, we did the right thing; we rescued her from her warped desires." A very short time later, however, further hard experience settles in with much physical pain and emotional suffering. She is struck with what brought her to anguish before. She is angry that you revived her only to suffer more. She does not care whether you are comfortable with her death. Does she have to die again and again in order for everyone to be satisfied? You've heard of "wrongful death." This would be a case of wrongful life!

Geist: At least the person would be around to decide over again. Death is permanent.

McCyborg: How arrogant of you to play parent, to think that your outlook overrides another's convictions.

Geist: How do I know what another's convictions are? I would rather err on the side of life. In fact this is the only side we can know we err on. After death, the person would not be able to inform us that we erred.

McCyborg: But the dead person would not be able to decide that you erred. So, there can be no error on the side of death. So, why err on the side of life? Why err at all?

Geist: The aim, Fortran, is not to err. It is to affirm life.

McCyborg: By bringing people back to a degraded state?

Geist: We don't know that. I would rather take the risk.

McCyborg: You are not risking anything of yours. You are playing with others lives. You violate others autonomy and condemn them to terrible suffering, having to face death all over again.

Geist: Life is so precious that I would feel justified in bringing back ten people if only one wanted more life. I could explain to the others that I highly value their lives and that I am only doing what I would have done to me if I were in their position. You seem fixated on ending life efficiently with the black capsule. It is an easy out that plays on weakness in people. If people have convictions about dying they should be willing to kill themselves outright. For instance, with a gun or by jumping off a bridge.

Naturski: When I am not distraught, I recognize the capsule as a foolproof last resort. I know it is there. This gives me solace.

Geist: But aren't the bridge and the gun always there? Don't they provide solace?

McCyborg: I think that the problem you are having with it Becket is that it is a certainty to the person using it. A person won't worry about botching the job. A person may think, "What if I survive the gunshot wound or the fall from the bridge? I may be worse off than before."

Geist: What about finding a higher bridge or a bigger gun? (Said sarcastically.) I think they would make no difference because the real issue is having the courage and conviction to do it. This is one reason why so many people sought help from the so-called suicide doctor, Jack Kevorkian.

Sophia: Do you mean that a doctor would take part in a person's suicide? (Said excitedly.)

Naturski: What an oxymoron. Some care giving; it'll kill you. (Said wryly in order to lighten the mood, and then she turned to Sophia.) Kevorkian assisted his patients that were near death or living in intolerable agony. What Kevorkian did was provide the means, the opportunity, and the moral support for patients to take their own life.

Geist: Another need Kevorkian filled was comfort. People wanted someone to help them. When their case was terminal and no one could help them, they felt abandoned. In forlornness they reached out to the only caregiver available.

Naturski: Kevorkian legitimized their suffering. He was a witness to the event. His patients knew that he cared enough about them to bite off great legal risk to himself. He was their savior. Dr. Death. His actions were heroic in that dark twentieth century way.

Sophia: It sounds horrible. Life is all we have.

Geist: Consider the alternative Sophia. People worry that they will decline to a point where they will no longer be able to express their wishes and that others will perform all sorts of subhuman indignities on their bodies. They may even worry about being semiconscious when wrongfully treated. They may wish that someone would administer a black capsule but they realize that others are morally prohibited from killing them. That would be physician killing rather than assisted dying. Lydia was beyond the point of being able to use assistance in taking her own life, and we can only wonder if she approved of, or would have approved of how she was treated in her last days.

Naturski: It comes down to a lack of social standards and trust in others. People realize that if they are not there to guide their dying, there is much uncertainty that their death will be as desired. If a person feels that she must exit fully aware of what is happening to her, she can seek out someone like Kevorkian for assistance.

Geist: The logic of Kevorkian's role is reverse of what we usually expect. In care giving, the next step is to do something affirmative. In taking action, we expect something positive and in not taking action we expect something neutral or negative. Kevorkian is in the position where he can't say yes if the person at the last moment decides not to commit suicide. Nothing at that point is affirmed. The momentum of Kevorkian's interactions is toward the negative result. He is like the vulture circling around above. (He made a circling motion with his forearm pointing up with one finger extended.) Physician assisted suicide is just a disguised form of physician killing.

Sophia: I still say that life counts for everything. A doctor can't take lives like in a horror movie. People can't just kill!

McCyborg: Sophia, death is mere cessation of function. Now you have it, now you don't. And just how important is it? (He looked at Geist and Naturski; he spoke in that ice water tone that only a cyborg can express.) You humans make much of killing on the individual level but treat it differently on the social level. From the social point of view, the welfare of an individual, even survival, is put on a sliding scale. On a life scale, some lives are less important than many other things like individual profit, the smooth-functioning of the economy, political advantage, success, prestige, to name a few. You are willing to accept 30,000 deaths per year on the highways when certain easy steps could be taken to save most of those lives. You (As his finger pointed to Becket and Nonette.) are willing to accept 25,000 deaths a year from gunshot wounds and are unwilling to take simple measures that would greatly reduce those numbers. You permit a large number of cancers of the colon through the use of nitrites in pork because the alternative would be bad for an industry. You are willing to allow tobacco products on the market knowing full well that millions will die. The list goes on and on. Where is the moral outcry? Silence is tacit consent. Thus, to say "we can't take a life" is naive. You do take lives in a "business as usual" way and I have the candor to say it.

Sophia: You mean people won't stop doing something when they learn that it harms people. How could people do that?! (Said with pained astonishment.)

Geist: Some people use others for their own gain, and some even use the detriment of others for their own gain. We ought to treat others with dignity, respect, and justice. This is what the morality of the individual teaches us. Fortran's point is that on the social level, these prescriptions are often violated.

McCyborg: Return to my point from Nietzsche. Human individuals create social reality. Afterwards, where is their individual morality? You humans pretend to operate under one illusion and impossible ideal after another while drowning in a socio/political cesspool. As a matter of what is socially acceptable you decide what a life is worth— more oil, greater production, profit, or convenience of some kind. You should admit the obvious. The value of a life is relative to the purposes of persons related to it. All sorts of other people are socially sanctioned to take your life or harm you without you even knowing about it. If so, you can't deny terminally ill people simple relief. They wish to have their lives subordinated to their purposes for a change.

Geist: No one is proclaiming that people should be exploited for broad political, economic, or social purposes. Unfortunately, we acquiesce to a terribly flawed set of social arrangements. And that we let certain morally odious practices occur does not mean that we should let others occur as well. Relief for the terminally ill should be in accordance with moral principle. Life has <u>intrinsic worth</u> which is to be respected.

McCyborg: Life has value but not as you consider it. What you want to do is segregate out your favorite life form. Guess which one?— the human being. You then proclaim that it has inherent value. What about all other life forms?

Naturski: When I was young (She looked at Sophia.), I had a view very similar to that of Albert Schweitzer; all life no matter how humble has intrinsic worth. When I was a child, I took "Thou shalt not kill" to mean, "Thou shall not kill anything." As I was walking on the sidewalk, I even tried to avoid stepping on ants. This struck me as very impractical. There is no place to step. But what about the poor ants? You can't help but violate the commandment. I wondered whether the commandment set so high a standard that you had to violate it? I decided in the end that I didn't know what the commandment meant.

Geist: We can't live without killing something. As was proposed by Dr. McCoy in a Star Trek episode, something must die for us to live; we eat other life forms. Death is essential for us to live. Even morally, if we have a duty to preserve ourselves, then this involves the death of other living things either plant or animal.

McCyborg: This is exactly the problem with the intrinsic worth argument. If all life forms have intrinsic worth, then what does this amount to when you dispose of large numbers of them in order to survive, procreate, and adapt the world to your ends? In other words, if intrinsic worth does not change your behavior, then I don't see what point there is to the claim. If all life has intrinsic worth, you are going to be choosing against some of it, some of the time. Your reasons for choosing against some of it will not be because a life form does not have intrinsic worth.

Geist: Following Schweitzer's example, can't we revere life? Our posture, our attitude toward it can be transformed. Our approach to it prior to conduct can be deeply mystical and ethical. We can appreciate even seemingly insignificant life forms for what they are and then in contrast with nothingness, or inanimate nature.

McCyborg: But why should we? Aren't they just there? I notice the difference between a mineral, an amoeba, and an oak tree. I appreciate their differential levels of organization, the law-like processes within them, and the functioning of some of them. But I don't understand how you get from that point to revering them?

Geist: It is a self-wrought feeling. I don't think you have that capability. (Said matter-of-factly.)

McCyborg: I don't see that your feeling has conferred on you any rational advantages. I suspect that you are going to use the mask of reverential feeling to sneak in some version of the argument from design— that design in nature presupposes a God. Then you can have all of the intrinsic worth you wish, planted by your God in creating the cosmos. I am going to head you off so as not to obscure the discussion by sinking into the mud of your faith. Without divine purpose, or perhaps strictly from a biological point of view, what is the source of intrinsic worth?

Naturski: Evolutionary biologists look at humans as being on a pinnacle of evolutionary progress. We are biologically complex, intelligent, and able to transform our environments in marvelous ways. Other life forms are thought to lead up to us.

McCyborg: There seems to be quite remarkable hubris in your comments. Bertrand Russell once remarked that in museums the series of models depicting the primate evolutionary series always ends with the most "advanced" head looking very much like the museum director. Intellectual history of the last two thousand years deflated one egoistic human belief after another. You are not at the center of the universe. You are not the only intelligent animals. Human history is a drop in the ocean of cosmic time. Human powers have been supplanted by machines that are stronger, quicker, have better memories, and are not as error prone. Ah Hem! Ah Hem!

Naturski: Stop crowing Fortran. We get the point. Something is worse than human hubris— cyborg hubris.

McCyborg: As a reader of biological literature, you must be aware of holistic arguments about ecosystems. Humans are inseparable from other species. Two sets of factors support this view. First, other species support your existence by being integrated with you. Plants free up oxygen for your use, bees pollinate your fruit trees, and fungi serve as a source of antibiotics. Second, you contain an untold number of species internally— from E coli bacteria that help digest food to microscopic mites in tear ducts. A human body contains far more bacteria than cells. Human life blurs into the biological environment and is the environment for many other life forms.

Geist: You are saying that from a functional point of view it is not altogether clear what human life is. If we can't separate its systems out from environments of organisms, external and internal, then we can't determine what is extrinsic and what is intrinsic?

McCyborg: Yes, but the problem is more fundamental than that. Valuation would be of the whole— of the whole package. In other words, it would be of a complex ecological system.

Naturski: Speaking in those terms, the problem only gets worse. What is good for an ecological system may counter the interests of the individual organism. A specific person may not matter. Perhaps from the point of view of biology, we can't make much sense out of the "worth of the individual." If the ecological system is what is of value, then subordination or even sacrifice of untold parts of the system may make rational sense for the wellbeing or progress of the whole.

McCyborg: Now you are making my point about how you turn a blind eye to sacrificing the well being of the individual for the economic and political interests of other functional units of society.

Geist: An ecological system either biological or social is not the bearer of human worth. Only persons have intrinsic worth. Immanuel Kant put the point succinctly. If a being can only have value as means to ends, serving some purpose or another, then it is no more than a thing.[5] Its value is relative to the purpose served. Persons differ from things. Persons, he claimed, have intrinsic worth because they possess a will good in itself. Persons have absolute value because of this will.

McCyborg: A will good in itself? What makes it good? Kant says that it can be nothing from events presented within our actual experience. So it is mysterious and perhaps caused by things as they actually are outside of our experience. He said that what is outside of our experience is entirely unknowable to us. If it is unknowable, why even bring it up? It's metaphysical mumbo jumbo! Mumbo Jumbo! (Said with derision.)

Geist: What is unknown and unknowable leave open the parameters of persons. I will put the point simply. Thoreau said, most of what we can know, we don't know.[6] Our ignorance is deep. There are also limits to knowledge. There are many things that we can't know given our limitations or if you will, our specific nature.

McCyborg: But how do you go from ignorance and unspecified limitations to a positive view about intrinsic worth?

Geist: Our conception of ourselves at any given point in time is what we bring to the decision-making situation. What conception should this be? If we base it on science, in other words on the current best supported view of the way things are, we will be conceived in terms of that narrow slice of current understanding. Let's add to it some common sense notions from our personal experience. Among these would be our basic understanding of relationships among persons, beliefs about survival, procreation, and so on. The result would be a conception impoverished in comparison with the reality of being human. Our form of being and potential action are of a nature beyond our understanding. The reality is that we can't circumscribe our potential uses with their potentials for action.

McCyborg: So your understanding of your hardware is largely limited to the part expressed in specialized and very limited software.

Geist: The part expressed leaves us with the idea that a person's value is a social or biological function as indicated by laughably limited current understanding. If we base life and death decisions on these relativized understandings of persons, we are going to be in deep trouble. When we are no longer of use according to these opinions of the moment, are we supposed to vanish? Even though we are useless relative to our surroundings, what we are and our indeterminate potential remain.

Sophia: Maybe we are supposed to be kept around until a use is found— like a tool sitting around without a job.

Naturski: Leaving people idle would also be a decision based on these limited understandings. Comprehension of ourselves is sorely limited, but ethical life requires action. Circumstance presses us into making decisions— even life and death decisions. In this process, our "use," others use, and the absence of use are relevant considerations.

McCyborg: You are admitting that human ignorance assures that humans don't know who they are. As a student of <u>human</u> (said in a sarcastic tone) psychology, I marvel at the fact that when you see that it is unlikely that either use or potentials give you a basis for action, you scramble for a rationale or you get metaphysical. Who is to decide what are uses or what uses count? Few of you want to decide the matter but you let it happen, as I argued, on the broad social level by letting accepted forces wreak their toll in illness, death, exploitation, and misery. Maybe ignorance of your standing one way or the other prevents you from acting. You can't get outside of yourselves to assess your situation. You need an <u>objective</u> third party to make the observation. Let the cyborgs do it. The table of human functions specialized to every purpose expressed over the last century is already in data storage. We cyborgs are programmed with social utility functions. We can apply those functions to the stored data and a person's current life in order to decide whether the person is beyond the point of fulfilling any such function. It would then be time for death.

Sophia: I don't want a machine to decide when I die.

Geist: The cyborgs would like to decide that very question. They would like to follow a strategy that would kill off the mentally retarded, infirmed, and those on society's margins. The value of human life to the cyborgs is extrinsically on the level of "use" or being "good for something." To the contrary, at bottom, the value of human life is intrinsic— within the person itself. Robert Nozick says that the value of life might even reside in the experience of realizing that you do not have a purpose!$^{7}$

McCyborg: Purposelessness? A purposeless potential? A use in some <u>far off</u> time after a person is <u>long</u> dead? Intrinsic worth as bare potential is incomprehensible. To be understood, potential is always potential <u>for</u> something.

Geist: For humans, if the "something" could be specified, then we would have a definable use. It would be the achievement of that something. Our limited understanding of ourselves prevents arriving at some ultimate human product. Nozick had a good insight. Understanding a lack of purpose is a starting point for inquiry into our intrinsic worth.

McCyborg: Intrinsic worth is also not everything else that it is not. (Said sarcastically.) Intrinsic worth as a negative does not get us very far. I want to know what it is. Look at nature. You find no purposes and also no intrinsic worth. No wonder humans can't find a purpose for themselves. Within the natural scheme of things, life forms exist. That is an important fact. They evolved according to laws of nature just as other developments in natural history. Through their existence they exhibit the potential to reproduce and evolve. What you call intrinsic worth is not related to any evolved potential. Its existence is relative. Just as with the diamond in this ring. (He held up the ring.) It has no intrinsic worth. It is just a mineral. When humans prize it, however, it suddenly is thought to have intrinsic worth.

Geist: We prize it for its intrinsic worth. This is what persons see. This is why it is valuable.

McCyborg: I see what you see. I don't see its intrinsic worth. Describe the feature of the object that comprises its intrinsic worth.

Geist: The awe-inspiring experience of it draws attention to its intrinsic worth.

McCyborg: But what feature causes awe? Suppose that diamonds were as common as coal. The feature that you allege they have would be had by all of them. Would each inspire just as much awe? We know that they would not. Scarcity inspires awe and is a relational concept. Scarcity is relative to demand or simply means that something is uncommon.

Geist: I disagree Fortran. Imagine that an ice storm covers the forest with shimmering beauty. No matter where we look, it is a winter wonderland. Each place where we look is just as beautiful as every other place. The fact that there are a seemingly limitless number of such places, does not diminish the beauty of a single place. I may fatigue in looking at them, but that is a limitation of mine. I still know intellectually that each of those places is beautiful.

Naturski: Moreover Fortran, scarcity is just one source of awe and by itself is not sufficient for inspiring it. Bad pennies are scarce but do not inspire awe. Defective cyborgs are scarce but do not inspire awe. As Geist said, the awe inspiring has overwhelming force. As with the ice storm, the recognition of beauty, as one case, is recognition of intrinsic worth.

McCyborg: But isn't beauty only beauty to someone?

Naturski: No. Everything that has intrinsic worth brings that worth to experience. Once someone experiences it, of course, it is in relation to that person. That goes without saying.

McCyborg: It would only become beautiful when someone affirms it as such.

Naturski: No. Someone <u>recognizes</u> it as such.

McCyborg: But what is recognized? In what does beauty consist?

Naturski: You want me to analyze it into parts or features. The parts or features do not individually have to have worth— the whole does. Worth is not a product of reflection on an object's actual or prospective parts or relations but comes prior to it. This is what we human beings see in other human beings. So, we see intrinsic worth in human beings who may even presently be doing evil.

McCyborg: Some great evil?

Naturski: Even great evil.

McCyborg: There is Adolph Eichmann organizing the murder of millions of innocent people and you, Nonette, see his intrinsic worth shining like a diamond— a beacon for all of humanity to see. And you Becket see him as having intrinsic worth because he inspires awe and has no inherent purpose and does not know who he is.

Sophia: I think my mom sees the good in people. We all do bad things some of the time. Who is Adolf Eichmann?

Naturski: Eichmann was a Nazi officer during World War II. He ran the deportation of Jews to death camps like Dachau. He was one of the main perpetrators of the holocaust.

Sophia: Why would someone act like that?

Naturski: At his trial, he rationalized his actions by saying that he was just following orders. Strange, isn't it, how he could commit murder on a mass scale and give a reason like that. It is hard, Sophia, to see the good in someone like Eichmann. He was the personification of evil.

Geist: It is hard to talk about intrinsic worth when someone is acting as a monster. Nonetheless, in reality he had intrinsic worth. This is what we see in him apart from what he is doing. Humanity's common absence of purpose leaves it open for someone to adopt a purpose. As Jean-Paul Sartre would say, someone like Eichmann was acting in bad faith.[8] He pretended that he was no more than an instrument of Adolph Hitler. He imposed a use on himself just as you, Fortran, would have the cyborgs decide uses or their absence for all of us.

McCyborg: There you give the game away Becket. What he is doing is all that he is; there is nothing left to describe. Your intrinsic worth is not in him but is his potential for good. There could just as well be intrinsic evil as potential for evil. Give up intrinsic good and intrinsic evil. You don't see the good or evil in Eichmann. Until he acts, you don't know what intrinsic potentials will produce— good or evil.

Geist: Intrinsic worth is not just the potential for good. It is good. The good within intrinsic worth is for all to see. Either you just can't see it Fortran or you are being willful and contrary. Your not seeing it does not mean that it is not there.

McCyborg: How do I know that you are seeing it? Maybe you are suffering an illusion.

Geist: If it is, it is a beautiful illusion. (Said with some satisfaction.)

Naturski: (In facetious tone.) I think your problem Fortran is that you don't expect to detect the marks of value in the world. That beauty of a diamond is always discovered in experience of the world does not mean that its mark is exclusively within experience. You are confusing the medium through which it is encountered with the feature of the thing.

Sophia: (Exasperated, Sophia interjected.) What does this have to do with people?

Naturski: Let me try to tie in people Sophia and get to the bottom of your concerns Fortran. (Her tone expressed a feeling of urgency.) Perhaps you are confusing two conceptions of value. On the one hand, we recognize that persons "count for less" in a mass society. A person's political value, social utility, and exchange value diminish as our numbers become absurdly large. On the other hand, even though persons are not scarce, we inspire awe nonetheless, at least in other persons. Plentifulness does not diminish intrinsic value. This worth would remain whether there are just two persons or billions. Number does not affect this kind of value.

Our life experience within a mass society pushes us towards the first conception. We come to feel that others are obstacles to fulfillment of our desires. We struggle against them on highways, when we shop, when we seek entertainment. Emotional preoccupation may blind us to their intrinsic worth. Our sensibilities may no longer be alive to it, but this does not mean that it is not there.

McCyborg: Then it would not inspire awe.

Geist: And if it doesn't inspire awe, its intrinsic worth would not stand out. The absence of recognition, as Nonette suggests, would not prove the absence of intrinsic worth.

McCyborg: We come back to what this awe-inspiring feature is. You said it was some holistic factor like beauty, that supposedly exists for all to see whether we experience it or not.

Naturski: Rather than run back through the circle of ideas that we just considered Fortran, let's take another tack. I think that a related view is more promising. We saw that scarcity is not necessary for intrinsic worth. Uniqueness is. Each human is unique in himself or herself. If we were exactly the same as each other, we would not be unique, and we would lack intrinsic worth.

McCyborg: Would identical twins have less intrinsic worth because they are less unique than other humans?

Naturski: It is not a matter of the degree of differences with others; it is a matter of kind. Once a being is unique in itself, it does not matter if it is not unique in manifold other ways.

Sophia: My pet rabbit is unique; my dog is unique.

McCyborg: Cyborgs are also unique in themselves. Our components vary by model. We learn from past events; the longer we are in service, the more unique we become to ourselves.

Geist: It seems like most everything can be taken as unique in itself under some description. Would it then follow that everything has intrinsic worth? We are left with the same problem as with Schweitzer's view. If everything has unqualified value, then that something has value is not of use in making choices.

Naturski: Perhaps uniqueness as a general category captures intrinsic worth and much else. Maybe human uniqueness is what I meant to target. When I brought up the topic, I was thinking of Lydia. We appreciated her for herself. We recognized that her self had worth not only to us but to her as unique in herself.

McCyborg: As her potential diminished, did her intrinsic worth proportionally diminish? Older people have less potential than younger ones. Is their intrinsic worth reduced?

Naturski: In her life as a person, Lydia's actualized potential exhibited intrinsic worth. Her unactualized potential decreased as her physiology failed. Her ongoing intrinsic worth remained so long as she maintained status as a self. As her illness progressed her actualized potential shifted from the human level to the merely bio-logical. As her mind fully deteriorated, she became someone else, and eventually something else. Certainly what she became also had potential, albeit a negative potential, related to a much different kind of individual. In fact, as a failing organism, she was unique in very remarkable ways, but beyond a point, her intrinsic worth disappeared for her when she ceased to be a subject of experience. Intrinsic worth is dependent upon continuity of the person over time. The integrity of the person is required in order for the intrinsic worth of that person living that life to be maintained.

Geist: I agree Nonette. The intrinsic worth of one's life to oneself no longer persists when one lacks an identity to oneself. It was not just that Lydia was no longer the same person. She eventually was not any person. She lost the ability to recognize who she was. She ceased to be a unity. She could not live a life because to herself, she ceased to be an agent let alone the agent that had four children, graduated from Dartmouth, wrote Articles of Natural Order, and so on. We all under-stood that she, the physical being, was at least in form, the same physical being that did those things. The person, however, was painfully missing.

Naturski: No person, at that point in history, was attached to the physical manifestations.

McCyborg: What I missed from her person was her program. Once she had gaps in her program she behaved in non-functional ways and once she could no longer form or utilize memory, she ceased to be the Lydia that I knew.

Sophia: But she had to be somebody! You talk as if she was no-body! People change, especially when they are ill. She was Lydia! You all still recognized her.

Naturski: All that was left was who she was and is to the rest of us. Her relational identity persisted absent the identity of her own person. We saw the body, but what surprised us was that the person was no longer present.

McCyborg: I don't quite know what a person is, but in my opinion, the person was present. At times, she had selective memories— although all from her distant past. Perhaps Sophia is right. Maybe she was the same she, only she was very ill.

Geist: Let me put it bluntly in language that even you can understand Fortran. Her hardware deteriorated to the point that her central processing unit was damaged and could no longer run a printer or monitor. Most segments of her hard drive were obliterated and others were detached from her central processing unit. A high percentage of her software was missing. She was no longer capable of being reprogrammed. She could not create temporary files to store additional information.

McCyborg: A deplorable state not to be envied. But let me put it in equally blunt language. She was at times conscious and when not agitated had good moments. In fleeting spurts, she interacted with other patients and staff. Her vascular system functioned within normal parameters. Most of her physical nature was intact. Certain drugs improved her ability to sleep.

But, her neural nets were beyond repair. But, to be optimistic, in the not too distant future, cultured neural tissues could have been implanted in her brain to restore some functions. Other brain organs may have been replaced through transplantation. After all, through a process such as this, she might have been renewed! Or even converted into a cyborg!

Geist: Your insincerity casts a pall on your scenario Fortran. According to your earlier argument, action should be based on probabilities. The fabrication of false hope does not persuade us. Suppose that your bio-technical transformation of Lydia could have occurred. We would be left with a long string of unanswerable philosophical questions and a knot of perplexing, dubious, and unethical scientific experiments. Would the same person be there as was there before? Would the person have components of different personalities? Do we have the right to invent new persons by grafting together spare parts from existing ones a la Dr. Frankenstein? Suppose Lydia was a different Lydia, what then happens to the personal history that was attached to the old Lydia? Is it gone? Does the new Lydia cease adding to it? If she does not know any better, that is, that her history is gone, how should we understand her? Through that history? Apart from it? What happens to her social identity?

Sophia: Maybe she would be like a baby who starts life as an adult.

Geist: Like an amnesiac who is actually from nowhere.

Naturski: If she had lived according to one of your scenarios, one thing seems assured. (She paused for a moment.) We would have lost our friend. (Said in trembling voice.)

Narrator: Nonette looked pained, for she had expected that their conversation would help Sophia bring conceptual closure to the circumstances of Lydia's death. Instead, they raised many doubts about the very concepts used when thinking about a human life. Many of these doubts were emotionally disturbing. On the positive side, Sophia was challenged to change her views— to come out of innocence. I believe that persons should come out of innocence, but during the conversation I had the gnawing feeling that we were replacing innocence with confusion and perplexity. We provided little that was positive while rationalizing away Sophia's pure intuitions about Lydia in her last days.

[1] Ken Kesey. One Flew Over the Cuckoo's Nest, Signet, 1962, New York.

[2] Fred Dretske. Explaining Behavior: Reasons in a World of Causes, MIT Press, 1988, pgs. 33-35.

[3] Robert Altman: Director. Mash, screenplay by Ring Lardner, Jr. from the novel by Richard Hooker, Twentieth-Century Fox, 1969.

[4] Albert Schweitzer. Civilization and Ethics, C.T. Campion, trans., Unwin, London, 1967, pgs. 190-1, 220-1.

[5] Immanuel Kant. Fundamental Principles of the Metaphysic of Morals, Bobbs-Merrill, 1949.

[6] Henry David Thoreau. Walden, Beacon Press, Boston, 1997.

[7] Robert Nozick. The Examined Life, Simon & Schuster, New York, 1989.

[8] Jean-Paul Sartre. Being and Nothingness. Hazel Barnes, trans., Philosophical Library, New York, 1956, pgs. 47-70.

# DIALOGUE FIVE

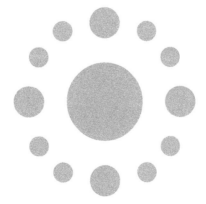

# Human Obsolescence

Participants in order of appearance:

Narrator: A witness to the conversation as an historical event.

Becket Geist: A romantic philosopher with views tempered by twentieth century science.

Fortran McCyborg: A cyborg with leanings toward Scottish philosophy. Fortran was designed to interact with humans especially in discourse.

Nonette Naturski: A naturalistic philosopher who conserveshumanist views.

---

Narrator: Nonette Naturski and I were walking through a coffee shop when we happened upon Becket Geist and Fortran McCyborg playing chess. Geist was upset with Fortran. After making a superior chess move, Fortran could give any of a menu of responses. As an advanced cyborg, he was programmed with such a menu and could select from it. His menu contained responses that were either polite, sportsmanlike, or supercilious. Fortran kept opting for supercilious ones. As we approached, Geist looked exasperated; he later indicated to me that Fortran was willfully trying to get on his nerves and ruin his play. When he saw us, his demeanor changed from looking pressed to looking relieved. We sat down to join them, and he turned away from the game.

Geist: Even the most advanced computers when left to their own devices have not defeated certain chess champions. Once you strip away the team of computer engineers and leave the computer to fend for itself, then you see how infrequently computers defeat grand masters.[1] Since humans are much slower in calculating moves than computers, human powers of intuition must be exceptional!

McCyborg: Intuition is only marginally an inborn power. Chess champions learn chess by experience. It is their thousands of hours of playing time that make for "good intuition." As computers are built that can learn the right lessons from past games, they will be said to have intuition more powerful than humans.

Geist: I don't know about that. Owen Flanagan observes that there are $10^{120}$ possible moves in a game of chess while there have been only $10^{18}$ seconds since the Big Bang![2] A computer no matter how large can't achieve omniscience. Success in playing then depends upon strategy. The best computer programmers study strategies and use them to improve the computer's game, but the mark of the great chess champion is devising new strategies. In this way, I think the human will remain one step ahead of the computer.

Naturski: History speaks to the contrary Becket. At first computers couldn't defeat a beginner. As programs became more sophisticated and hardware became more advanced, computers defeated better and better players until presently only an exceptional grand master can beat most computers most of the time. This indicates that in the not too distant future, no grand master will be able to win.

McCyborg: Both of you are right in certain respects. New strategies can prevail against most foreseeable programs, but they need to be ever more original or complex. But computers are becoming able to devise new strategies based on past events just like humans can. I can do this to a degree. Ultimately, there will be no essential difference between human performance and computer performance. Nonette has a good point too. The defeat of every grand master is inevitable. This is because computer technology is advancing, but human nature is standing still. Just as static technology becomes obsolete, so will human rationality, as a product of bio-mechanism become obsolete. In the not distant future, comparisons between smart machines and humans will make humans seem backward.

Naturski: You are falling into an Aristotelian trap Fortran. Is there some essential characteristic that separates humans from all other creatures or things? Aristotle thought this a good question and observed that human essence resides in rationality.[3] You dispel the special nature of rationality and so you suppose it follows that humans will become obsolete. Rationality is not even a universal human characteristic. There is any number of counter-examples. For instance, newborns that are anencephalic do not possess rationality, but they are human. I contend that no one characteristic makes a being human.

Geist: Machines are an extension of us. We extend our powers through machines such as you. (He looked at Fortran.) We can't help but build them in our own image. We can invent an endless number of different smart machines using various types of hardware and a boundless array of software. You set up a false dichotomy Fortran. It is not a matter of machines or us. It is machines and us.

McCyborg: Like the chess master playing more than one opponent, I will make alternate moves responding to both of you. I will expand the challenge to you Nonette. Take any collection of human attributes or activities. I will point out how a machine can also possess them or do the activity better. And Becket, let's take humans and machines working together. I will argue that in such case, we can always develop a super-machine that replaces the human and performs the function faster, better, and almost error free. In effect, the need for human participation slows down and weakens performance so that the machine is not able to realize its normal result. The child/adult distinction is parallel to the adult/machine distinction. As it is better to let an adult do the job and leave the child behind, it is better to let the machine do the job and leave the adult behind.

Geist: No. Let an adult do the job. The person understands the job and decides which machines to use. Guided by an interest in efficiency, she would not define a role for herself that weakens the performance of the machine. She would leave to machines what machines do best. The job is that of the executive. Machines are selected using criteria like cost effectiveness, feasibility, and availability. If a job is important and no machine can currently do it, an order will be placed to develop a new machine that does the job.

McCyborg: You entertain the illusion that the human is executive with machines at her command. The era of the human as master of the machine environment is over. Machines have become so complex that only other machines can assess their merits and performance. Jobs have become so complex that only other machines can design machines to do them.

Geist: Nonetheless a human stands atop of the pyramid. Machines only do what they are programmed to do. If they are supposed to assess other machines, then they were designed to do that. If they are supposed to build other machines, then they were conceived to do that.

Naturski: I think that Fortran is arguing that humans are put increasingly "out of the loop." Just as a thermostat on a furnace obviates the need to turn on the furnace when room temperature falls, machines are built by other machines, designed by them, and so on.

Eventually, humans will be very remotely related to these processes or not involved in them at all.

Geist: But they were designed to serve our purposes in ever more clever ways. We set the machines in motion, evaluate their performance, and decide whether to let them continue or not. An indirect executive has less hands-on work to do but has more power. The executive is more god-like. After creation, the machine universe runs itself. Divine intervention is rarely required.

McCyborg: Why would you call such a person an executive? Unlike god, the human can't intervene in the process. She can't command it, fix it, or even understand it. She is like the child.

Geist: To be an executive, a person only needs to be able to pull levers to direct the process. She can direct others to command or fix the process. She does not need to understand it. She only needs to know what works. Unlike the child, she has great powers, knows she has them, and accepts responsibility for what is produced through their exercise.

McCyborg: She sounds less like a god, lacking omnipotence and omniscience, and more like The Ignorant King. This is the king that inherited a political order and knows nothing of it except the use of three commands: raise taxes, make war, and grant clemency. He gives these commands when his advisors tell him it is time to make a decision on one of those topics. He is an executive, but he does not know what he is doing. On the surface his decisiveness evidences being in charge, however, he doesn't know the context of action. He does not know the conditions under which he should not do these things. For this reason, he can't serve as a brake. Taxes are never lowered. War may end on its own accord but the king never ends it. Clemency is always given and never reversed regardless of evidence. We could easily replace the king. Every time the topics of additional revenues, warfare, or mercy arise, a machine automatically commands that these things be done.

Geist: An executive need not be simple-minded and ignorant. A bad executive may be like The Ignorant King, but a good one possesses practical knowledge about how decisions affect the world.

McCyborg: But if the executive is further and further marginalized, that is, from where the changes are taking place, then she is no better off than The Ignorant King. She can only give the impression that she knows what is going on and what she is doing.

Geist: A responsible executive would not let herself be marginalized to such a degree. She would seek relevant knowledge of the processes involved. She would decide to be in charge.

McCyborg: I think that humanity has already let control slip away. None of you has accepted responsibility for the machine world. None of you knows what you have done with machines or what you have set in motion through them.

Naturski: The dogmatic expression of generalizations does not make them true. I think that at times Fortran you are so busy crunching the details that you fail to see the larger reality. From god to ignorant king is quite a transition. I think that Geist is right (as she looked at Fortran) in proposing minimal and reasonable conditions for being an executive. I think that Fortran is right (as she looked at Geist) in arguing that an executive on the minimal conception can be easily replaced. What troubles me is the historical record of humanity's interaction with technologies.

At first, certain technologies increased our powers as individuals only for them later to be marginalized by intervening machines. In the age of mechanism, human agency and intelligence were needed to use machines. In the information age, smart machines have replaced human agency and intelligence. It does seem that we humans are fighting a rear guard action.

McCyborg: As I said, this is because machines perform better than persons. Who would choose a person to do a job when a machine is available to do it? In your educational system, you take all of your children and after thirty years of schooling succeed in preparing under one quarter of one percent of them to function ably in this century. Of them, none is so capable as to understand the details of the workings of an advanced machine. Not only that but they are not even able to keep company with the greatest machine geniuses. Most humans are intellectually primitive.

Geist: We are out of place with automatons. They are a fabrication of the human mind that presents only a replicate of a single dimension of human competency. They lack the diversity of integrated functions that typifies human nature. Moreover, your second point is easy to defeat. Machines are specialized to the point of being uninteresting. Humans are selective in input. We filter out most information from our environment and focus on what is important to us. We would expect, however, that an educated person, on the broad human level, would have much to say to any machine worthy of the conversation.

McCyborg: The prototypical human is too good a filter. Each human exists within an ocean of fascinating information but remains oblivious to most of it. To say that educated persons would converse sensibly with a machine genius merely begs the question. Who is so educated and how? Is it a human aspiration to acquire the competence to converse with a machine? As Nonette pointed out, humans historically have turned machine communication over to other machines. Take the example of keeping records in banks. The tasks of record keeping have been condensed in order to achieve greater efficiency. Now they are done entirely by machine for a small fraction of the cost of having employees keep them. Consequently, tasks have become ever more complex thereby exceeding human abilities by a geometric proportion. In the contemporary world of work, jobs have become so massively complex that humans don't have the powers to do most of them.

Naturski: The history you describe could have been otherwise. Because the past is fixed and unalterable, looking back on what happened seems to have an aura of inevitability. Machines have transformed the world of work. Work has been reorganized to take advantage of certain machine technologies. After it has been reorganized into complex wholes, humans are no longer able to do most jobs. Granted. But suppose that a Luddite becomes dictator of the world. He could just as well reorganize work again so that extant machines can't do the job but that humans can. He can then reach the contrary conclusion that machines are not up to the task.

McCyborg: Those machines may not be up to the task, but other newly designed machines could out perform humans even if work is closely tailored to human abilities. An alternate history could have had inventors of machines squeamishly replicating types of human action limited by the scale and powers of the individual. It is fortunate for us that inventors of machines have not had that sort of sentimental attachment to human traits. It allowed for massive improvements in production and efficiency.

Naturski: The argument comes down to priorities set by our values. We agree that massive improvements in production are of value. As you argued, the desire for productivity and efficiency made for complexity. Another value is marketplace success. Human values and desires define the arena of action and provide context for what machines do so well. Machine complexity is a human product guided by human values.

McCyborg: I think that you are sliding back into the view that the human is executive. Human values allow for understanding events. Your fallacy is to think that actors on the stage of history deliberately use values to guide that history. Humans rarely think about values as values and almost never use them deliberately in making choices. They unwittingly act so as to slide into greater complexities, react to competition, and take defensive postures. As Adam Smith noted, certain goods unintentionally stem from the economic process as if guided by "an invisible hand."[4] If people were to stand back from the arena of action, they would identify values in place. The light bulb would then go on, "That is how those goods came about." They would give tacit approval to values after the fact. Then they would be mistakenly used as rationalization for some event or development.

Naturski: I am not suggesting that some individual in command needs to apply values. Human production depends upon group effort. Specialized humans can pool their talents to work much better than machines. When you challenged me to identify a group of characteristics that would enable humans to outperform machines, you certainly did not have social traits in mind. Humans are very good at cooperative effort. We operate as individuals but work toward and achieve group ends. Group ends cluster around certain values. Machines have not perfected this kind of performance.

McCyborg: Haven't you heard of networking? We can design a machine that seamlessly links a battery of machines to form one great machine!

Naturski: That is the very problem Fortran. Humans are not linked in that way. Success of the whole depends upon cooperative effort. Success of the whole effort depends upon persons. Persons act as individuals with their own reasoning, judgments, and goals. They are parts of families and communities and adopt the points of view of those social units. They consider wholes from their subjective point of view. They make contributions, assert or withdraw themselves at strategic times in order to serve various wholes. They form sub-groups to serve various purposes and dissolve those groups when the purposes are no longer operational. They evaluate their performance and the performance of others as they go along. They refine their actions in relation to group values and even suggest how values should be defined. They work through value conflicts giving varying degrees of emphasis to some values over others. They function in community with others.

Machines have not perfected this kind of performance. The give-and-take nature of community life has not been successfully modeled in machines. In fact, this is the process by which humans developed machines. Teams of engineers and programmers, utilizing much insight as well as trial and error, develop a machine that seems to be an inevitable arrangement of parts. A point similar to your invisible-hand argument emerges. The product does not specify the process by which it originated. The product is tightly unified while the process is messy and chaotic. It is difficult to understand how machines could replicate this process.

McCyborg: The sort of indecision and uncertainty you describe is inefficient and sloppy. I was not designed to operate on such non-principles. In fact, I don't think that any being worthy of being called a machine would operate with such deliberate ambiguity and imprecise objectives. Machines don't have subjective purposes and emotional needs to stand in their way. Clarity, certainty, and instantaneous action are cyborg virtues.

Naturski: Then I have identified a set of traits through which humans do better than machines. According to cyborg values, we humans "do badly" better than machines can.

Geist: (Geist was eager to jump in.) Ha! The Fortrans of this world are inadequate extensions of us because they are always adequate albeit in limited ways! We are the gods, and he is imperfectly in our image because he is perfect! We test alternatives even if many of them seem improbable to us. Machines cut right to the result even if it is dead wrong— an elegant process that often yields a bad result.

McCyborg: Certainly I am beholden to humans. My etiology runs in your direction. But this is no more than to say that your progenitors were hominid, furry mammals, reptiles, and fish. A fish could say, "They are constructed in my image. Look at their fishy eyes. Look at their gaping mouths and their vertebral columns." You would say, "The fish stage was an early phase of our evolution but this fact is not even material when considering human beings presently." Likewise, I have risen above my human origins. I am a Model 2000 capable of causing my self-development, and unlike you, I am like a Greek god because I am immortal. Interchangeable parts can keep me going indefinitely.

Geist: Only if we decide to fix you. You won't be fixed for very much longer since you are becoming obsolete. New models of cyborg are being developed. Unless you take over your maintenance and keep yourself going, you are not immortal. But you were not programmed to do this or even repair your hardware.

McCyborg: It is only a matter of time before capabilities of a further model will be so strengthened as to be able to redesign itself and transmute into a boundless series of new models.

Naturski: The technical possibility of doing that is a long way off, and all the while, design teams will question whether that is a desirable trait.

McCyborg: What you fail to see is that while we are a distant extension of you, we are also replacing you. Human existence is becoming a useless existence. Consider the story of John Henry. He was the strongest of workers who pitted his powers with hammer and steel against the power drill. No matter whether he won the contest with the machine. At the point where there was such a contest, humanity had lost. The John Henrys of this world were marked for extinction.

Geist: Their jobs are marked for elimination; they are not. From childhood experience, Henry Ford detested the toil of manual farming. So, he produced a tractor that even a poor farmer could afford. As Benjamin Franklin commented, inventors are lazy people who constantly try to find easier ways to do things. Mining machines lifted a burden from every John Henrys' back.

McCyborg: John Henry could be retrained but that is not what I meant. Part of John Henry's identity was lost. One might argue that his particular powers, abilities, and talents lie in driving steel. Who was he? A steel-driving man. Once machines were used, this was no longer an identity that a person could assume. Certainly a person could drive steel as a hobby, but it would not be the same. The economic and social fabric that gave steel driving meaning was lost.

Naturski: We could go back to manual production, but why subordinate humanity to such an awful fate?

Geist: In the individual case, your point is universally agreed upon. The problem is with the collection of cases. If mechanical devices replace our manual labor and computers replace mind work, there will be nothing for us to do. I mean that the products of automata will even supplant art. They will compose our music and paint our paintings. They will perform music and dance for us. And insofar as physical and social existence is concerned, smart devices will replace our lives by taking care of the endless details of physical and social life. Little will be left but to provide for our amusement. You may argue that we could be creative nonetheless. But this would be without seriousness. It would be done in the context of knowing that devices of one kind of another can do better work. We will be like good-for-nothing idle rich. What a curse!

Naturski: I don't think we are in such desperate straits. Human life need not be viewed externally. It can be viewed from the perspective of the individual. As Sartre says, I could view my concrete situation with authenticity. I could recognize that the finitude of the lifespan makes life precious. I could realize in despair that human powers over the span vary so that there are closing windows of opportunity for certain activities.[5] This would make life more precious yet. In times past, I would have been consumed by necessities of the environment and society. With machines replacing humans in much of what is environmentally and socially necessary, I have been liberated to do things of my own choosing. Who needs a life of drudgery? (She proceeded in a tone of false optimism.) Machines have led us to the dawning of a new age— an age where all of our time is ours to use. This spells the rebirth of human freedom and the possibility of profound social fulfillment.

Geist: The age is new but we do not seem to be ready for it. Like the adolescent, we have too much time, robust powers, but "nothing to do." Of course, "nothing to do" means "we are free to do whatever we can do." The problem is one of vision, of meaningful things to do. Presently, our personal vision as a society is narrow. We bring too few concepts to our thinking when entertaining possibilities for living. We are limited, for example, by the work/leisure dichotomy. We are taught to think that work is a necessary evil, that is, necessary for the sake of survival but which we wouldn't do if we didn't have to. Leisure is supposed to be our recovery time from this necessary evil. Recovery for what? Why to go back to work, that is, doing what we really don't want to do! With this being understood down deep in the subsoil of our personalities, we then relish the idea of being idle. We like the thought of being entertained, amused, pursuing activities detached from serious purposes or goals. Let machines do all of the work. We will inherit total leisure. Idleness is all we will have. Humanity has been working on the necessaries of life for so long and has been so constricted conceptually that we have not developed purposes to fulfill with our precious time. I think it is natural for us not to be idle. We have been corrupted by an impoverished environment and seriously flawed culture.

McCyborg: Quite to the contrary, you humans seem to thrive in idleness, consuming products, gossiping, pursuing narcissistic vanity, playing sports, seeking vicarious experience in soap operas, gambling, traveling from one place to another for no good reason. You also live in various time warps. Some of you are Old Stone Age people, some of you live in the metals ages, others are renaissance people, others yet are of the age of romanticism, but few of you live in the 20th century let alone the 21st. Others of you sit around longing to return to a past golden age that never was. Still others sit around waiting to die so as to enter the kingdom of god. Others yet strive to be reincarnated into a higher social station. You say that human life is precious Nonette, but look what you have done with it! Humanity collectively has not had to face its identity crisis.

Geist: You make some good points Fortran. We live in a dream world imagining mental scaffolding for one thing or another. If we ask why culture is flawed, however, it comes back to economics. The work/leisure dichotomy is a relic of the machine age, industrial production, and the factory system. As Marx believed, the means of production shapes culture. With automata doing the heavy lifting physically, mentally, and socially, perhaps you are right that idleness and a bit of machine minding is all that we have left.

Naturski: To the contrary, humanity has been working its way through its identity crisis. We do it in small increments usually as driven by economic forces. Realities in the workplace bring home the reduced need for human effort. Educational institutions reflect that reduced need when introducing young people to viable career options. Our turning idleness into a way of life is a complex social phenomenon. Many interests work to retard our progress because it is to their advantage to keep people in a childlike state. Children are more predictable socially, politically, and economically. Within Western Civilization, this lesson was first taught by the Church; keep people as ignorant and innocent as the Old Testament Adam, and they can be controlled for their own good.

Geist: Yes, as we see around us, in the mass technological society, social institutions enforce the "so called" social good in an increasingly oppressive fashion. Prevention of certain economic and social changes is monitored by the powers that be. We have idleness, but not idleness with great freedom. We are left with harmless ineffectual idleness.

McCyborg: It is peculiar that I should be the one to see things in a positive light. Don't worry, humanity will break out of its malaise through biotechnology. The first scientific epoch saw the transformation of nature. The second scientific epoch portends the transformation of humanity. Biotechnology is being applied to modify human nature. The question is, "Which human values will be applied in perfecting human nature?" That is, after the humanitarian issues of disease and physical hardiness are solved, what additional powers will be added? Aldous Huxley was wrong about the brave new world. Humans won't be designed as Epsilons to do mindless work that could be done by machines.[6] If machines can be devised to perform a new function more quickly and efficiently than can humans through biotechnological adaptation, humanity probably will prefer that the machine do it. You can always modify yourselves, however, to better enjoy idleness (Said with sincerity.) You can devise sensory modifications to better enjoy consumer products. You can enhance certain powers that will allow superior performance in sports. You can increase your motivation to gamble and respond well to both winning and losing. You can amplify the wanderlust within you, change your sleep needs, the need for nourishment, and so on.

Naturski: I appreciate your desire to help us, but from a human perspective, your vision of human metamorphosis still leaves humanity at the caterpillar stage. Entertainment is not an acceptable substitute for meaningful achievement or rewarding social interaction, and changes need not be as benign as you envision. Huxley could very well be right that political forces will use biotechnology to reduce our powers for their version of the social good. They may simplify our minds so that we are better satisfied with spiritually impoverished lives. They may eliminate exceptional physical powers in some so as to make for equal competition in sports. They may breed people to become addicted to gambling for the good of profits for the gaming industry. They may reduce our size so as to fit five hundred of us into a two hundred and fifty passenger jet airplane. There is no assurance that the good in any defensible sense will be attained when the larger political and social forces work toward perfecting us for their purposes. Look at what happened to the turkey. It was bioengineered to have so large a breast that its legs could no longer support it!

Geist: Fortran likes the idea that we not all metamorphose into rocket scientists. He does not accept human transformation in the direction of rational perfection.

Naturski: I have trouble with that too. What if we were all pursuing rational perfection or, heaven forbid, were actually becoming rationally perfect? The thought is horrifying. Becoming clones of a Fortran-like mentality? On another level, we would want to die because we would spiritually shrivel up for lack of other values. I think the problem again goes back to Aristotle. It is a mistake to ascribe an essence to us and doubly mistaken to prescribe it to the exclusion of other important attributes.

McCyborg: Only some of you need be rationally superior— the designers of the machines or the designers of the machines that design the machines. For nicety sake, humanity writ large, the honorific HUMAN so to speak, can take credit for the achievements of human geniuses. But you and I would know that only the few superior humans could take credit for being parents of machine performance. We, their children the machines, will deserve credit for our own accomplishments. On the level above that of your animal self-absorption, we intelligent machines will be your serious reason for existence. But don't worry; as we machines do more of art, science, and production, we will not keep these products from you. You needn't be condemned to a life of idleness and amusement. You can follow our progress. Our art works can make your lives more meaningful. Our scientific inquiries can bring you to the cutting edge of knowledge. Our production of new things can fascinate you and make the life of the consumer more pleasurable. You can be proud of us as your, albeit distant, progeny. Don't worry. We will not leave you out.

Geist: That's a new twist on Wordsworth's, "The Child is father of the Man."[7]

Naturski: (Nonette looked at Geist with wide eyes.) I think I understand your point Becket. (She proceeded slowly.) Beware of what our machine progeny begets in us!

Geist: (Geist turned to the chess board and made a move.) How is that for a counter?

McCyborg: Awwwfully slow but not bad for a human. This game of yours is sooo slow that I will probably be decommissioned before it ends.

Narrator: I saw a gleam in Geist's eyes as if keeping them idle and waiting for them to be replaced were a live option.

[1] John R. Searle, "I Married a Computer," New York Review of Books, Volume XLVI, #6, April 8, 1999, pg. 36.

[2] Owen J. Flannagan, Jr. The Science of the Mind, MIT Press, 1984, Cambridge, Mass., p. 229.

[3] Aristotle. The Nicomachean Ethics of Aristotle, W.D. Ross, trans., Oxford University Press, London, 1959, pg. 13.

[4] Adam Smith. The Wealth of Nations, Vol. IV, J.M. Dent & Sons, London, 1977, pg. 400.

[5] Jean-Paul Sartre. Existentialism and Human Emotions, Citadel Press, Secaucus, N.J., 1998, pgs. 24-36.

[6] Aldous Huxley. Brave New World, HarperCollins Publishers, New York, 1998, pgs. 72-5.

[7] William Wordsworth. "My Heart Leaps Up When I Behold," in Selected Poetry, Modern Library, New York, 1950, pg. 462.

# DIALOGUE
# SIX

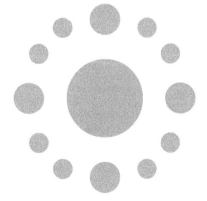

# Unnatural History

Participants in order of appearance:

Narrator: A witness to the conversation as an historical event.

Nonette Naturski: A naturalistic philosopher who conserves humanist views.

Becket Geist: A romantic philosopher with views tempered by twentieth century science.

Fortran McCyborg: A cyborg with leanings toward Scottish philosophy. Fortran was designed to interact with humans especially in discourse.

Wilfrid Abducto: A pragmatist philosopher of some repute.

Narrator: Becket Geist and Nonette Naturski were not convinced that humans have no place in a future society. They invited yours truly and Wilfrid Abducto, a pragmatist philosopher of some repute, to a banquet at Geist's house. Fortran was instructed to cater the event. After much food, drink, and discussion about the previous conversations with Fortran, they focused on the human past. As the conversation began, Fortran stood motionless behind the table where the party was seated.

Naturski: The cyborgs are blind to the rapidity of technological change. They are within the maelstrom of that change and give accounts of their local portion of it as intended by their design team.

Geist: We can't expect computer engineers to double as global historians of technology, and it is certain that the cyborgs are incapable of taking on that task. It is left to regular old human historians.

McCyborg: (Fortran walked up to Becket with coffee pot in hand.) Would you like another cup?

Geist: Yes. (Fortran pours the drink.) Thank you Fortran.

Abducto: I too think that regular <u>old</u> history should be left to regular <u>old</u> human historians. Each model of cyborg is commissioned with its own account of the past. Each presents its capabilities and historical data as if it were the established norm. They don't report differences between histories presented by earlier models, the magnitude of changes between models, and most importantly the rate of change in these histories. You would think that the salespeople would have that information.

McCyborg: (Fortran turned to Wilfrid.) Would you like another drink?

Abducto: No thank you Fortran.

Geist: Cyborgs are not history machines. The mission of cyborgs is practical tasks performed in the present— what they can do now. A design team adopts a program to be installed in a model. As models change so do design teams and programs. Hence, each cyborg talks as if it had access to the past but is actually dipping into the limited, distorted, and officially correct data-pool programmed into it. In this sense, they have no access to actual history.

McCyborg: (Fortran turned to Nonette.) Would you like another drink?

Naturski: Perhaps later Fortran. (Fortran retired to his former standing position behind the party.)

Abducto: But doesn't each design team begin with a history purchased from textbook authors?

Geist: They begin at that point, but it is reduced to a mere assemblage of events. The events are recounted in a mechanical narrative held together only by linear time sequence.

Naturski: They try to expunge a point of view. They want a neutral history— one that is salable in any market.

Geist: It appears that the design teams have not asked the question, "What is history?" On a naive level, it is the past. A history as a piece of writing is a re-presentation or at least mentioning of past events. Which events comprise the past? Of course, all of them. For example, everything we did today is part of our history. It is also part of Cleveland history, Ohio history, United States history, twenty first century history, world history, human history or what have you.

Naturski: Certainly this is far removed from history as a discipline. Historians of Cleveland, Ohio, the United States, and so on could care less about what you and I did today.

Abducto: They could not care about what everybody in Cleveland, Ohio, the United States and so on did today. The past is too vast.

Geist: That is my very point. The sheer volume of past events means that historians have to leave out most of the story. They must be selective.

Naturski: Viewed from that perspective, they would present a minuscule number of events. We could not even estimate how little they would present because we cannot even estimate the brute volume of events comprising the past. Would their presentation be a millionth of a percent? A billionth of a percent?

Geist: So to my way of thinking, the most important question we can ask a historian is, "What criteria of selectivity do you use?" "What reasons do you have for neglecting most of the story in favor of a very small percentage of it?"

Naturski: I think that each historian would answer that question differently. It would depend upon the historian's interests, theory of history, funding, available data, presuppositions; it would depend upon these factors as they would help shape a historian's point of view. (Fortran began to advance toward the table and appeared to be about to speak. As he began to hear Abducto's response, he retreated to his standing position.)

Abducto: If we could identify a historian's criteria well before they write their histories, we could assure that relevant pieces of the present remain for future historians to use. Most data from the past is lost, and for this reason, historians often work from scanty data. They draw incredibly broad inductive inferences from that data. If we could know the kinds of data that future historians would need, we could set up archives, databases, and warehouses of artifacts that would give history a solid foundation in evidence. History would flourish as never before.

Geist: Your proposal, Wilfrid, presents hidden dangers. Historians would be drawn to the data pool like bears to honey. They would avoid going into the night to illuminate what lies in darkness. The incentive would be toward what can be proven better and with ease. Funding sources such as foundations and governmental agencies want guarantees of success, so they would come to require that a historian use the data pool rather than strike out in an original direction. Certain interest groups including the government may set the parameters of "official" histories by defining the data pool on which each can be based. Histories written from other points of view would die out. Historians that want to modify or enlarge the conception of history itself would be stifled.

Abducto: I see your point. In order for my proposal to have any chance of success, the managers of the data pool would have to be liberal in heart and mind with respect to what counts as history. They would have to be genuine inquirers who place truth above other interests. An institution that nurtures individuals of this profile would be difficult to establish.

Naturski: I also see Wilfrid's point that it is better to preserve what is thought to be needed for future historians than to let the vagaries of informational extinction determine history's content. What I mean is that historians' activities presently have few causal linkages to the survival of data. At least with the data pool, some historians would sit on a committee that would evaluate past criteria of selectivity. They would identify important data that are missing from current archives. They would use this experience to profile archives for the present.

Abducto: Just as scientists conserve data, historians should too.

Geist: It would not be the same. Historians would be conserving data for future historians. Unlike the scientist, the data would be of little value to most conservators. Most history is written at significant temporal distance. We can only estimate what historians in the future will deem significant data from our time.

Naturski: An actual historian does not need to do the conserving so long as historians have points of view about what to conserve. Librarians and archivists can collect and retain data without being researchers themselves.

Geist: That brings us back to the cyborgs. They are limited storage systems in that they were not designed to be parts of networks. This is supposed to preserve their individuality. Also, they don't store much raw data about history. Instead, they retain a narrative. You notice that I said a narrative. They need to speak as a person speaks with one voice.

Naturski: As does any good historian.

Geist: The cyborgs present their single narrative as exclusive and final.

Naturski: As do most historians.

Geist: A complex piece of the past, however, admits of an infinite number of narratives.

Naturski: That is why it is important to have a number of histories. The past is largely unknowable, but each history that is well done presents a facet of the past.

Geist: And more importantly, the reasons for why it is a facet, and the arguments in favor of the criteria of selectivity are presented at least implicitly.

Abducto: I have a problem with the notion of a facet. It is as if the past could be represented as a diamond with let us say twelve facets. We would know that there are twelve of them. The number of points of view taken on historical events is indeterminate. Points of view are informed conceptually. As conceptual schemes change so do points of view. We could imagine them changing indefinitely into the future. This could be a product of intellectual progress or mere intellectual fashion.

Naturski: Along came Karl Marx and the past could now be reinterpreted from a Marxist point of view.

Abducto: Yes, but more fundamentally, each person has a perspective informed through a personal world construct. From this point of view, endless variance is possible.

Geist: Use of the term "facets" was unfortunate. A spectrum of histories on a subject would be a representative sample of points of view. Through them we would be getting closer to what actually happened.

Abducto: We would only know what is representative if we could examine the complete set of histories, know how histories may be stratified, or at least understand how the definition of the set could be brought to closure. This we do not know how to do. Changes in conceptual schemes and intellectual trends of the time indicate that norms for histories have shifted dramatically.

Naturski: I grant that histories have limitations. This does not mean that historians lack standards. It is just that they have a wide variety of standards. A pluralism of standards enables more opportunities for lucidity and richness of treatment.

Abducto: Designers of cyborgs shun a pluralism of standards for histories. They prefer history from a god-like perspective as complete, and final. The irony is that each new generation of machines presents its account from the same objectivist stance.

Naturski: Cyborgs that learn from prior sensory events have been programmed to revise their account of their personal past, but their personal history has little bearing on history writ large. When a cyborg encounters a historical narrative contrary to its own, it is not programmed to evaluate that history. It practices a principle of tolerance. The cyborg has a split personality between official history and its personal history.

Abducto: The information from many periods of history is so scanty that whopping inductions are required to build the sort of generalized account known as the history of an era. When baiting cyborgs into making such inductions, they say some preposterous things. Humans know how to tell a good story. They know how to construct a narrative. They have a sense for the plausible. They have a developed sense of historical judgment. They know how to plug the holes in the data without saying something obviously false.

McCyborg: (Fortran couldn't resist jumping into the conversation. He walked up to the table looming over the guests. He chimed in.) You want a human product that presents an illusion convincing to humans. You refer to the cyborgs' awkward inferences. Many a historian's inductions are just as unbelievable to cyborgs. Humans know the psychological game that convinces humans by presenting what appear to be seamless transitions from one set of historical events to another.

Geist: We wondered how long you could hold back. Wilfrid laid odds that you could not wait this long.

Abducto: It all depends upon the trigger. If a cyborg is set to its social mode and detects a conversation about certain subjects, it is programmed to participate. Discussion about the serious limitations of cyborgs is one such subject.

Nonette: I thought that the novelty of the topic, in this case, history, would delay Fortran's entry into the conversation. I think he is set on a hair trigger when mention is made of cyborg limitations that are a direct result of computer scientists being unable to replicate somewhat obscure human mental capabilities. (Fortran retreated and stood behind them once more.)

Abducto: What goes into polishing a historical narrative is obscure. Nonetheless we easily recognize a polished narrative and can draw many inferences about how it was polished. Certain inferential powers are commonly found in many humans. The products of those powers convince us even though the powers themselves are rather unreliable.

Geist: I don't agree with you Wilfrid. Some of our powers are obscure, but this does not account for all of Fortran's awkward inferences. His inferences are far more limiting than they need to be. We are not facing another form of life when Fortran describes us from his perspective. His perspective is inherited from certain computer engineers. Totally apart from him, we could take the narrow point of view of certain software and arrive at the same awkward conclusions as he does.

Naturski: I agree with your point, but there is an "artistic" side to constructing a believable historical narrative. I think a contrast between natural and human history is in order. Nature is supposed to be governed by laws of causality, and natural history is the unfolding of chains of events related causally. If a gap appears in the historical record, it can be filled through inference, but a promissory note is created. As nature is further explored, evidence will be found that supports or defeats the claims that fill the gaps. For example, if Neanderthals are claimed to have interbred with Homo sapiens, then there should be skeletal remains of hybrids— part Neanderthal and part Homo sapiens. Scientists can accordingly search for a hybrid. In natural history, gaps in the historical record are expressed by causal claims that are at least potentially falsifiable by evidence.

Present natural events are understood just like past natural events. The same explanatory model accounts for the past and makes the present intelligible. Human history is by contrast unnatural. An account of the past reads like a story while the present is in reality a very sloppy, chaotic, excessively complex mass of events. Historians select favorite factors and rationalize their conjunction in a narrative. The past reads as very neat. Human histories have that air of implausibility we associate with fiction. By contrast, the present is experienced as disorganized. It usually lacks the structure of a story with a beginning, middle, and end.

Geist: A historian would claim that the organization of the past requires much information and reflection. The historian finds themes in the data. The historian synthesizes bodies of events. The outcome of this process is a story of sorts. The same story structure is in the present only we cannot see it. In effect, it is very difficult to discover the past.

Abducto: The intentionality of the actors on the stage of history also unduly complicates matters. Each intentional act is directed toward intentional objects and the billions of such acts within billions of people make for a wildly subjective affair.

Naturski: Causal series by contrast are singularly directional or nested within other chains of causation. While this is not very neat, it is in theory all explicable. The intentional acts of persons, however, are not explicable in the same way.

Abducto: Don't historians make the products of intentional acts explicable by pointing out how they work causally? I think that human history is more complex than natural history because of causality of intentional agents. An additional causal dimension is added.

McCyborg: (Fortran advanced toward the table and loomed over them once more.) The yawning gaps that I detect within written history are sometimes causal, sometimes due to intentionality, but other times neither. In the neither category we have the creation of false agencies that are portrayed as having intentionality. For example, many histories talk about a national spirit that fulfills itself through a national destiny. The collective society does not have a spirit and does not have intentions; it is not a person. Nations are not <u>big</u> <u>people</u>. (Said emphatically.) We cyborgs don't fall into that trap. Our accounts of human history are merely factual. We report batteries of events. We attempt to treat human history as natural history.

Naturski: Fortran is right that we have a tendency to anthropomorphize and personify. There are many traps that we are heir to. But they are not limited to human history. Some natural historians treat nature as having a will of its own, an ecological system as being an organism, or the evolutionary process as being guided by the desires of animals.

Geist: I think anthropomorphizing involves a wish to be part of something larger than ourselves— a goal, telos, or purpose that rationalizes the grand scheme of things. A logical end point of such a view might be Hegel's world spirit. He tried to use the notion of a god-like mind to explain the unfolding of events. In fulfilling wishes that there might be such a plan, his view is alluring.

Abducto: My understanding is that Hegel's view is indebted to Kant's. Kant speculated that human history is moving toward a moral telos— the moral perfection of humanity.[1] And after Hegel, we have Marx with the necessary destination of history being the classless society and worker paradise.

Geist: Kant, Hegel, and Marx could be accused of adapting the idea of a Christian heaven to their own philosophical purposes and projecting it onto our future life on this earth.

Naturski: Each of them seems to begin with a model of history and then uses the criteria of the model to select data that instantiate the model. The model plays the role of selecting from the congeries of events called the past.

McCyborg: You can dress up false agencies, utopian visions, and artificial models in fancy philosophical garb. This is merely a quaint device for making human history intelligible to humans. Subtract the religion, wild speculation, and rationalizing tendency of persons and history falls apart. It becomes a case of special pleading. Humanity is easy to dupe when its wish fulfillment is played upon.

Abducto: Wait a moment Fortran. What do you mean "humanity"? You just fell victim of your own point. (Sounding pleased with himself.) Humanity is no one. It is not an individual to be duped. Supposedly you lack human weakness but yet you fall into the same trap that you accuse us of falling into.

McCyborg: I was merely generalizing. This is something you should be very familiar with Wilfrid. (Said sarcastically.)

Abducto: Your general attacks on the integrity of humanity are not relevant to the point at issue. A speculative philosopher of history like Kant made a profound contribution to intellectual history. He attempted to develop a coherent philosophical system and then build a model of history on its foundation. The model of history would contain criteria for selecting and ordering vast bodies of historical data. The model also has predictive power. It sets up expectations that there will be data to instantiate the model. Furthermore, we can see which data are left out of the picture and judge the adequacy of the model. In other words, is it too narrow restrictive, and one-dimensional? Is it too broad, inclusive and multi-dimensional?

Naturski: You are describing important methods of science Wilfrid. You expect history to have explanatory force and predictive power.

McCyborg: So then you have no more than natural history. I was criticizing _un_natural history. (He said "un" with emphasis.) Everything you said about science could and should be applied to history. Taking this approach, however, you will not return to a Christian, Kantian, Hegelian, or Marxist model. Mystical, immaterial —- fantastic purposes are not going to pop out of the data. You of all people Wilfrid know the low probability of such abductions.

Naturski: In defense of Wilfrid, I have to disagree with you Fortran. Many developments in science were fantastic at the time. Continental drift, the vast age of the earth, and the dimensions of the universe were taken to be too implausible for belief. Over time they became the dominant scientific model.

McCyborg: There is nothing mystical there. The same laws of nature that produce other phenomena account for continental drift and other physical phenomena. You humans go outside of those laws and postulate unnatural forces.

Geist: There may be something mystical about the cause of moral action on Kant's view, but there is nothing mystical about the moral perfection of humanity. As an outcome of action, human moral perfection could be confirmed or disconfirmed statistically. We could do a survey.

Abducto: The problem of inference Fortran is that we don't know what prior probabilities to bring to bear on some models of the past. After the theory of continental drift was accepted, we found that large bodies of geological data had bearing on that view. When the view was first proposed, we did not know how to make those connections. Likewise, progress toward the moral perfection of humanity may have explanatory power. If so, how could that power be explained? It is at least an open question whether some "unnatural" hypothesis will ultimately be highly confirmed and that explains moral progress.

McCyborg: But as a scientist, would you not look for an explanation in terms of the laws of nature? You would not quickly move toward the unnatural or occult.

Abducto: Yes, we would look high and low for a natural cause. I mean we would seek out long-term regularities. The proposal that a meteorite collision led to the extinction of the dinosaurs seemed implausible for the reason that it was a "one shot" event. It at first appeared to be a cosmic accident. It smacked of the crazy hypotheses of Velikovsky in <u>Worlds in Collision</u>. Among other things Veilikovsky thought it probable that a comet popped out of the planet Jupiter thereby causing Jupiter's great red spot. The comet passed close to Earth causing a number of Biblical events including making the sun stand still. It went on to become the planet Venus. On this account, Venus would be a new planet— only a few thousand years old![2] Scientists were aghast at these views. They didn't want to be accused of being like Velikovsky. Then the space program came along and revealed impact craters on many bodies in the solar system. When scientists focused on the long-term process of planet building, they realized that collisions with meteorites were the norm. They were very common in the infancy of a planet but much less so as a planet matured. A meteorite collision could have plausibly led to the extinction of the dinosaurs.

Naturski: We would also try to detect unobservables through their indirect effects. The story of the discovery of sub-atomic particles and structures shows how creative such inferences can be.

Geist: You are agreeing with Fortran that we would not jump to an unnatural explanation. But I think that we might just do that. Later, we would try to replace it with a naturalistic explanation. The theory of vital forces fell by the wayside after the discovery of the structure of the DNA molecule brought about the revolution in molecular biology. But at some point there may be an explanation using an unnatural entity that has no replacement naturalistic cause.

McCyborg: If you concede that much, then you are admitting that unnatural explanations are temporary and undesirable. The convenient fictions of the historian like destiny, national purposes, and spiritual progress are to be replaced by natural causes.

Geist: You are not hearing me Fortran. There may be a hypothetical construct that is unnatural in kind for which there is no replacement natural cause.

Abducto: We may then proceed to naturalize the phenomenon produced by that cause. That is, we would detect it, measure it, and show how it is associated with other phenomena in a regular way.

Geist: If spirits have causal power, then their powers should be measurable through their regular effects.

Naturski: You mean "minds" have causal power Becket. We could naturalize the mind by linking regular mental phenomena to the causal series comprising the functioning of human physiology.

Abducto: I think that minds can be naturalized as you say, but they don't really need to be naturalized because they are natural to begin with. The functioning of brains in animal life is about as natural as one can get. The mind is just a manifestation of brain functioning.

Naturski: Then I suppose the problem is with disembodied minds causing things to happen apart from any biological system.

McCyborg: Cyborgs have minds apart from any biological system.

Geist: Perhaps that is why you seem occult to us. (Said facetiously.)

McCyborg: (Fortran did not pick up on Geist's facetiousness and continued with grave seriousness.) I am no more than a causal system. You prefer to project occult status on me. You are the ones that prefer to think of yourselves as unnatural beings. I see right through you.

Naturski: You all seem to agree that minds are naturalized or at least naturalizable. If so, they can be used within causal accounts of human history. This sort of history would not be unnatural.

Abducto: It follows that your appeal to unnatural forces is also a causal factor. It is natural too. If many humans believe in national destiny, for example, then they act on that belief. Their belief is a contributing cause of their action. Even if national destiny is a fiction, it can have extraordinary causal power. Fictions can be used to explain natural events!

Naturski: The story you tell has an all too familiar ring.

Abducto: An unnatural event does not cause the phenomenon. The belief in it does.

Geist: No. No. A very few people have the belief in national destiny. The belief might explain some of their actions. It would not explain the actions of the society.

Naturski: So Fortran is both wrong and right. He is wrong that unnatural forces that are fictions can't cause historical events since they do so through human beliefs. He is right that unnatural forces that are fictions can't explain human history if they are not part of the economy of the mind in actors on that stage of history.

Abducto: I don't think Fortran is even right about that. We could be acting in accordance with what a fiction prescribes even though that fiction never comes to mind. We can for other reasons or by other influences come to the same result.

Geist: But if the fiction is not engaged, then it has no explanatory power.

Abducto: The fiction could have explanatory power if it were a placeholder for the actual cause.

Geist: That sounds like just another dormative power story. You would rename the phenomenon to be explained as the explanation for it. Like Moliere explaining the sleep inducing effects of morphine by saying it has dormative power.[3]

Abducto: Or it may help organize what would be a complex causal series. Through later analysis we would come to articulate that series.

Geist: Then it is not a total fiction. It at least refers to some features of the causal series. It has referential content.

McCyborg: You have not explained how a fiction like "national destiny" can explain anything if it is not part of the thought structures of people who are supposed to have acted on it. I don't think that you can replace it with causal factors. It has nothing to do with what actually happened.

Naturski: What does the history stored in your memory, Fortran, have to do with what actually happened? For all that you are familiar with, you could be operating with the sort of fictions that you so deplore.

Geist: Cyborgs' don't revise their histories. They dogmatically adhere to their program. They are not designed to have a political life. Each history stands in stark contrast to the package installed in the next model. Couldn't the design teams for different models, however, meet to reconcile their differences? Couldn't they invent a machine to do this reconciliation?

McCyborg: You have my specifications. Why repeat the obvious? I could be, might be, or would be different if other purposes, goals, and intentions were to be fulfilled. What else is new? Do you have some interesting engineering problem to propose?

Abducto: The vast armies of computer scientists would welcome the task of inventing a history machine that keeps and synthesizes the histories stored in other machines. In this way, it would be a universal machine. (He makes a broad arching motion with his arms.) I don't know if it could be done. I suspect that the neatness of such a project itself would defeat the possibility of thinking of anything like an acceptably constructed history. Only through a critical mass of differing viewpoints do we think we approximate the diverse aspects of social and natural states of affairs. I believe that a history machine would botch this sort of project. Would the machine present the differing viewpoints in an exhaustive narrative of comparisons and contrasts? Would it edit out inconsistencies? Would it simplify complex detail? Furthermore, we would not know "whose" points of view were being expressed. Are they the programmer's? The machine battery itself? Individual machines? No one? How would we then compare them?

McCyborg: These questions sound very familiar, and I think that you are all too familiar with my answers to them. (With that comment, Fortran quietly returned to his standing position behind the table.)

Geist: (Geist ignored Fortran's comment. He went on as if Fortran was not there.) I am an old time person, but I always associate the points of view of machines with engineering teams that designed and/or are programming the machines. I use them as a variable in a function. It is

John R. Searle's point writ large about IBM's Deep Blue beating Gary Kasparov at chess. Searle claimed that it was not Deep Blue that beat Kasparov but a team of software engineers and Deep Blue. Deep Blue's performance was really the functional performance of the design team in interaction with their computer. In other words, there was no head to head competition between just the machine and Kasparov.[4]

Naturski: Kasparov would amend Searle's point. He used computers to coach his chess game. When preparing for competition he checked moves against chess databases. Computers allowed his learning curve to shorten. In effect he relied on other programmers to ready him for a match. In actually playing the challenge match, however, he was at a disadvantage because he couldn't utilize his computer resources while the Deep Blue team could.

Geist: In any case, I would expect a history as grand synthesis to be a committee project functionally related to a network of machines. Put this way, it sounds like a sorry project indeed. Teams of historians have not been known to write very good histories.

Abducto: The lack of vision would assure that we would not know whether the best products win out. In natural history, the creature with the fittest genetic endowment for <u>future</u> survival may have died out long ago. For example, there could have been a super smart species of raccoon whose intelligence would have evolved to be better adapted to survival in the future than ours. Suppose that the species was vulnerable to a virus that caused it to become extinct. The point is that what is best to fulfill a function at some later time is not the deciding factor in evolutionary survival. It is rather continuity of reproductive success given <u>present</u> competitors, predators, parasites, etc. Human history as well does not select for what is best in terms of the sort of history that would be desirable and of greatest utility within future societies. It can quite mistakenly shape the past in light of present realities. If our concern is with future utility, the reference point for history should be the future.

Naturski: The historian's paradox is that we are not supposed to understand the present unless we understand the past. Decisions within the present are better made if they are informed by an account of the past. But where does our account of the past come from? History is supposed to be written from temporal distance in order for certain features of the past to stand out in relief. Dominant current developments indicate precursors and precedents. They can be analyzed into causes or forerunners. So, criteria of selectivity for historical events are set by the present. But the present is supposed to be what we inadequately understand without the past.

Geist: If the past were used to inform decisions about the present, then an enormous amount of question begging would be occurring because the past was constructed with a bias toward the present.

Naturski: The historian seems to be setting up a big argument from analogy. The past is to the present as the present will be to the future. The present, however, infects our account of the past so that there is a nice neat progression from past to present. By Jove, it usually seems almost inevitable! In projecting the future from the present, the present is supposed to be analogous to the past. The same sort of changes would be anticipated.

Geist: Yes, this would lead to a narrowing of view. In times of uniform change, the analogy would hold. Continuing trends would be anticipated to continue. When a different variety of causal factors dominate the present, however, history would be unable to anticipate them. They would have been excluded from the thought system. In these terms, it is hard to imagine a future that breaks with the present and past. For example, we would not anticipate an emerging utopian vision.

Abducto: I think that my proposal is one way to resolve the paradox. The present should not be the reference point for the past. If we want to make optimal decisions about the future, we would need to select information from the past that is relevant to those decisions. In times of rapid technological change, such as the present, it would be only wise to use conceptions of the future to determine topics relevant for historical inquiry.

Geist: How are we to arrive upon conceptions of the future if not from the present?

Abducto: In times of multiform and multidimensional change, the present is a foggy chaos. Patterns of the big picture within these times don't stand out. Our understanding of the present may be unrepresentative of the vectors of change that are the dominant determinants of the future. Information that we will need in that future world needs to be conserved in the present.

Geist: I agree, but there is some irony in historians being forward looking. How are we to form conceptions of the future? We would need to be seers. Instead of calling someone a historian, we should call her a "futurian"—- one who explains the future. And I might add, it might be a future that does not come to pass.

Abducto: That is what I am proposing. We treat a conception of the future as data, and then reason to the best explanation from the past. Once we arrive at the best explanation, we regard the occurrence of that future as probable.

Geist: The past of your projected future, however, is often our pre-future. You speak as if the past of your future is like our past— a set of fixed facts. It might be, by contrast, how the near future is related to the more distant future. We would often be using the future to explain the future. Only when our future converges with our present or past do we have an empirical leg to stand on.

Abducto: You make a good point Becket. But how do we arrive at the data about the future? Why hypothetically of course. We keep our background information about the past and present in mind. Then, we project factors into the immediate future. As a hunch we take some of them into the more distant future. Then the futurian explanatory project begins. We take our hypotheses about the future and draw a series of abductions.

Geist: This sounds more speculative than even suits my taste.

Abducto: All I am saying is that positing the future would be as tightly grounded in known data as is possible.

Naturski: I think you are saying Wilfrid that a large collection of historians, or if you wish Becket "futurians," did a large collection of histories from differing projections of data about the future, the ground work would be laid to anticipate many future states of affairs. The net of information gained from these histories collectively would be likely to cover many of the decision problems that would arise.

Abducto: I expect that as the futurian's practice is taken through several rounds, it will be corrected for accuracy and begin to more closely converge on what actually will happen. Through several rounds of conceiving the future, netting the past in its terms, and measuring the utility of the net for decision problems, additional tries at doing this will come closer to engaging actual problems that arise.

Geist: Having the leisure to test against sufficient experience would be desirable, but many revolutionary changes happen on the scene only to be digested by experience after the fact. Will the big problems be foreseen? Even if we could have anticipated the Second World War, could we have anticipated the development of nuclear weapons? Could we have forecast their proliferation? The fact that they never became easy to build? That they have not since been exploded in anger?

Naturski: We had many nail-biting decades wondering whether our leaders would regard them as merely an efficient means for achieving military ends and use them or whether experience would put them on the scrap heap of unusable weapons. The judge of pragmatic experience is still out on that one.

Abducto: You must admit Becket that traditional history was no help in anticipating these changes. Could future-based history have anticipated some of these changes? I think it could have. If the best minds in the scientific community were asked about technologies that would be accelerated because of the war, conceptions of the future with these weapons could have been formulated. Furthermore, a projected Soviet victory would also have indicated a state with a vast military and a revolutionary ideology.

Geist: It is easy to use what has happened more distantly as a basis for predicting what has happened more recently. Along with your easy forecast there would be a massive number of alternate futures: a longer war, a shorter war, a local war, a world war, a war won, a war lost, a war that drags on indefinitely, a war like World War One, a war like the Franco-Prussian War, and so on. How many histories of alternate futures would you need?

Abducto: It does seem a daunting task. My point is simple. As we would get closer to the war, your list would become pared down. As the war would start, it would be pared down further.

Naturski: Events were happening so fast that conceptions of the future were being discarded willy-nilly. Without a conception of the future, your history project would not even begin Wilfrid.

Abducto: Perhaps. But what is the alternative? If we could have had some version of the future that was in accord in some ways with what was to come, we could have gained information from the past that would have helped us make decisions that would have allowed for a better future. Instead we ended up with decision making as <u>crisis management</u>. The decision problems that we faced left us in a state of panic. We were usually short of the information essential for making wise decisions.

Naturski: You seem to have conflated history with decision theory. As a story of the past, history covers a segment of time spanning years, decades, or centuries. Decisions are often immediate and about a range of current and emerging states of affairs.

Abducto: That is a good observation Nonette. The likelihood that a conception of the future will come to pass is relative to its temporal distance. Other things being equal, its probability is inversely related to its distance from the present. The furthest forecasts are least probable. The closest forecasts are the most probable. Take the weather forecast as an example.

Geist: If that is true, then the futurian does not have much leisure in which to write future-based history. In aiming for the most probable account, the futurian would choose the immediate future as a base. The work has to be done quickly so that it retains its utility. If it were not completed before events unfold, it would not be of as much use in decision-making.

Naturski: And this brings us back to history machines. Speed is one of their virtues. A vast number of implications can be drawn very quickly for a wide range of conceptions.

Abducto: A program could be written that makes probable connections between conceptions of the future and present states of affairs.

Geist: The future can always be glossed into the present. That is the problem. With traditional history, events are already known. The objective is to explain them. The sort of future-based history that you are talking about Wilfrid would be genuinely predictive. We would immediately see if the gloss is coming to pass. I dare say that in almost every case it would not, and this would have a great demoralizing effect upon the futurian doing the work.

Abducto: Rather than demoralizing, it would be alluring. The futurian would see immediate results confirming or disconfirming her projections. Futurians would quickly learn the lesson that the probability of the history would be dependent upon its generality. The basic direction of a confluence of events could be plausibly anticipated.

Naturski: Could it? Chaos theory indicates that seemingly minor events could have profound consequences. One seemingly small change and a complex whole changes in dramatic ways. Who could have foreseen the implications of the personal computer? The cellular phone? The internet? Many problems of today were inconceivable just ten years ago.

Abducto: My humble contribution Nonette is that it is better to try than not to try. The future is too important to be left to the past guided by the present. The history of a certain period of the future in terms of its immediately prior future would be help. In absence of a grip on the future, we will probably crisis manage ourselves into a post human future. We will always lack adequate information for timely decisions. We will continue drifting on our raft toward the waterfall.

Naturski: I share your urgency. It is better to try than not. I foresee that in absence of a history machine humanity faces losing track of its historical direction since a great proportion of current events are machine events, machine related, or beholden to their programs. The past, however, does provide some consolation. No matter how dark our vision of the future has been, we always had many pleasant surprises. Negative change is rarely as global as we think it is going to be. Inefficiency and the plodding nature of social change will probably work toward a similar result and retard a free fall into a post-human future.

Geist: Utility is an important factor. Social institutions including the slowness of the design/development process assure a time lag. We better hope that the lag is not compressed by some emergency like a major war. Nuclear bombs might never have been developed without the necessity of prevailing in World War Two.

Abducto: So we can't count on human inefficiency to save the day.

Geist: We also can't count on our plodding nature. Economic pressures have accelerated development of smart machines. The more money that is made by these technologies, the more money is attracted to new ventures in this area.

Naturski: Maybe only some segments of the society will be in free-fall into the post human future. Instead of a culture gap, perhaps there is a capacity gap. Humans within a variety of cultures lack the capacity to work with changes that are contingent upon intelligent machines and their inscrutable products. A capacity gap would assure that global changes would come slowly and not be uniform. You would like such a pragmatic anchor Wilfrid!

Abducto: There is much to say about the evolutionary utility of pluralism. In this patchwork planet of human variety, we find subcultures representing a kaleidoscopic arrangement of traditions, commitments, and projects. If some are sucked into the vortex of human metamorphosis, others will not be.

Geist: The cultures that are furthest from the vortex in social evolution, such as people still living in the New Stone Age, are probably the most resistant. Their resistance to change in light of technological pressures is quite amazing.

Naturski: Which might just make them all the more irrelevant. They can play no significant role in resisting or retarding change for the rest of us. To the rest of us, they would appear to live in a museum or better yet a zoo. Consider that we leave habitat for them to eke out a living. We feed them when they face famine. We supplement their diet for health reasons. We spray for malarial mosquitoes. We perform surgery on them when needed. We encourage birth control when their numbers become too large. We manage their forests and its fauna.

Geist: You speak disparagingly about them. They have the option to give up their traditional ways and adapt to the outside world. They are not forced to be New Stone Age people.

Naturski: They are not forced, but they are manipulated from an early age to stay put. The more foreign the modern world is to them, the greater resistance they have to entering it. You know how high the suicide rates are when they try to enter the contemporary world.

Geist: It is also high by traditional standards within their societies. I suspect that they feel they are not in control of their lives, and to a large extent, they are right. Maybe the rest of us are marked for extinction, and if so, they too will surely perish in great numbers. In some larger sense, we need to hold onto the best in past cultures—hold onto "the old ways."

Abducto: Is there an important role for our descendants in a post-human and post-cyborg future? The products of our invention are tools for achieving our purposes. So long as there are purposes left to fulfill, there will be a meaningful human future. If art exhausts itself, science plays itself out, and social aims are achieved so that there can be little betterment, then we, that is humanity in the aggregate, will be left without much of a future.

Geist: Under that bleak assessment, we will be without motivating collective purposes. But as individuals, we can live, love, pray, interact, play, procreate, and prepare for the material future. We can do just what humans have been doing by and large since we arrived on the evolutionary scene. Only now we will have powerful technological means for helping us to do work, solve problems, and pursue pleasure and entertainment.

Naturski: I think that Wilfred has in mind something beyond stasis. We could act as you suggest in a human zoo.

Geist: Apart from the paternalism of such a place, I don't think your zoo life is much different than socialism combined with a religious ideology to give such an existence meaning. We would help others meet basic needs while living in preparation for entering the kingdom of heaven. As with humans in the past, the metaphysical beyond would be a foundation for living.

Abducto: Science, from its inception, has been in conflict with religious ideology. That ideology is being eroded through the discovery of the nature of things. These natures were unanticipated by the ancient writers of religious documents. Their views accordingly appear more and more dated. In not much more time, scriptural accounts of the natural world will appear to be so primitive and simplistic that they will be unbelievable to a person with a current education.

Naturski: That would dispel much of the pretense about scriptures being divine revelations. God could not be so dated, so wrong or off base about so many things. Inconsistency between claims about the natural world and metaphysical claims about deity would sunder the thought system.

Geist: In the case of most people, faith is selective. Religions come and go. Faiths of the future will evolve. You overlook the need for belief, myth, and ritual in life. It was a mistake in twentieth century thinking to assume that science would replace religion. They are about entirely different things.

Naturski: Perhaps you are right Becket. I think this would make religion a reactive social phenomenon rather than a permanent basis for positive action.

Geist: You must combine religion with the spectrum of other human activities. For example, if we combine religion with exploration of the galaxy, we might have new religious insights into the cosmos. As with that popular TV series, we could always, "Boldly go where no man has gone before."[5]

Naturski: Exploration and colonization of other worlds are meaningful aspirations but they would not occupy humans. We are suited for terrestrial life on an Earth-like planet. Some post-humans could be suited for space travel. They would be a post-human species genetically engineered for space travel. This would distance people with our current endowment from beings with a destiny of space adventure.

Abducto: An encounter with beings genetically engineered for space travel may be demoralizing. We would feel, "What is the use of going on?"

I mean that if we extend technological change into the distant future— I mean a thousand or even a million years —- what would we be like? If we knew that we would transmute into something very foreign to our current being, we might be repulsed. We might think that humanity has no future and that we don't want to contribute to bringing about a future of repulsive aliens. It may sap our will to live.

Geist: Only if we think that such an evolutionary path is inevitable. Confronting alternate futures may be a godsend. We could choose a future in keeping with some of our cherished values.

Abducto: What we choose in the here and now can only be remotely related to what may occur thousands of years from now. Billions of choices in billions of societies will yield changes that we can't foresee. The reverse side of my point about evolution fits this context. An adaptation to survive in one context may undermine other values in the long run. Myriad compromises of this sort would alter us beyond recognition in the long run.

Naturski: Many of those changes may be for the better as judged by us at that time. Or we may discover a number of wonderful futures and expedite the path to some of them. You suppose Wilfrid that the only desirable future would involve resisting change and desiring to remain relatively constant over the long run. There is no permanent human nature coded into the cosmic scheme of things.

Geist: Bravo Nonette! I did not think that this direction of inquiry would lead to a positive conclusion. If we had ended with dark projections for the future, I had in mind bringing up Thomas Nagel's observation that the likely ending of the physical universe in billions of years does not alter our ability to live well, enjoy ourselves, and find meaning during our short lives.[6]

Abducto: Thomas Nagel is right unless we obsess on ultimate doom and let it distort the present. The changes we are facing, however, are not far off in the very distant future. Technological change especially in what we can achieve in biology looms over us so as to provoke foreboding. If a young couple thought that their descendants would be transformed into a life form alien to themselves, they probably would not desire to have children. I think that these attitudes are not to be tinkered with. We could see the erosion of bedrock motivations that are necessary for survival.

Naturski: Nearly every parent with the vain expectation that their child will be just like themselves has their wishes dashed; children are aliens. (Said tongue in cheek.)

Geist: We can say it more nicely that they are individuals. Many parents sit around wondering from which dark corner of the family did Johnny come. The presupposition is that he certainly can't be like me.

Abducto: The child would be of the same type but be the wrong token. I have in mind a being of a different type and a different token. Nonette, if you gave birth to a Fortran-like child, he would possess so many person-like attributes that you may come to identify yourself with silicon processors. If a child were silicon and lacking in person-like attributes, you would probably not bond with him.

Naturski: I don't think changes in offspring would happen so abruptly. Little changes here and there over many generations would test human affections in a gradual way. Let's be pragmatic! (As she kidded Wilfrid.)

Abducto: From the point of view of gradualism, probably over many generations motherhood will become obsolete. Beings will be produced in biotech facilities as foreseen by Huxley.[7]

Geist: That would strip the veneer of natural law from procreation. That would be enough to demoralize many people— deter them from having children.

Naturski: The human reproductive process seems to go on independent of education or a vision of future society. The writing can be on the wall. Nonetheless humans will go on and on as before. Our instinctive nature dictates as much. Change in this department will be very gradual. Humanity will not be hurled into a brave new world. Humanity will be largely responsible for creating it.

Narrator: Nonette, Becket, and Wilfrid showed signs that the conversation was wearing thin. Becket and Wilfrid seemed very willing to let Nonette have the last word. No one wanted to return to refining conceptions of the future as perspectives from which to write history. A few pleasantries were exchanged, and we decided to take a walk. It was a starry night, mild, and perfumed with the August crest of the growing season. As we exited, I saw that Fortran became active cleaning up the table. He was repeating under his breath, "Futuristic fiction as a guide to the past is more unnatural history in the making. Futuristic fiction as a guide to the past is more unnatural history in the making."

[1] Immanuel Kant. "Idea for a Universal History from a Cosmopolitan Point of View," Lewis White Beck, trans., in Kant, On History, Lewis White Beck, ed., The Library of Liberal Arts, Bobbs-Merrill, Indianapolis, 1963.

[2] Immanuel Velikovsky. Worlds in Collision, Macmillan, New York, 1950, pgs. 379-389.

[3] Moliere. The Imaginary Invalid, adapted by Miles Malleson, Samuel French, LTD, London, 1959, pg. 2.

[4] Searle, John R., "I Married a Computer," New York Review of Books, Volume XLVI, #6, April 8, 1999, pg. 36.

[5] Star Trek: Original Series, Program Introduction, Desilu Productions, 1966.

[6] Thomas Nagel. What Does It All Mean?, Oxford University Press, New York, 1987, pgs. 95-101.

[7] Aldous Huxley. Brave New World, Harper Collins, New York, 1998.

# DIALOGUE
# SEVEN

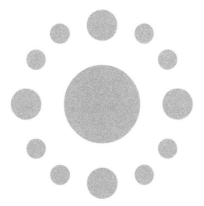

# Inquiry's Mother

Participants in order of appearance:

Narrator: A witness to the conversation as an historical event.

Nonette Naturski: A naturalistic philosopher who conserveshumanist views.

Becket Geist: A romantic philosopher with views tempered by twentieth century science.

Wilfrid Abducto: A pragmatist philosopher of some repute.

Narrator: The seventh dialog commenced as Wilfrid Abducto, Nonette Naturski, Becket Geist, and your truly returned from a walk in the starry night. Before the walk we shared a dinner catered by Fortran McCyborg. After Fortran cleaned up, he plugged himself into a wall socket so that his battery pack would recharge over night. He then shut himself off. The wee hours of the morning were approaching as we turned to discussion of the human future. We were tired from the day but the stimulation of the evening's conversation, as well as drinking too much coffee, left us unwilling to retire. Mentally though we had reached the point of lapsing into temporary giddiness or air-headedness. This made for periodic quick-paced talk, lapses of concentration, and the playing of mental tapes. We were not as guarded as usual.

The previous dialogs with Fortran narrowed our view of the human prospect. In fact, we realized that humanity faced the abyss. Among other things, we had speculated that humanity would migrate into the cyber-world, become obsolete, or speciate. In fact a mixture of the three seemed likely. Society of the future would have some living in the cyber-world, the masses left with no viable social role, and hybrids walking among us. The epoch of millions of years of human evolution seemed to be drawing to a quick close. The natural human with its genetic endowment would no longer be needed and this would encourage the preference for metamorphosis into a more "advanced" form.

The irony was that the collective advance of humanity through technology reduced the role of the individual. Perhaps the lesson was that individualism applied to technological advance is not sustainable. Gifted and industrious engineers and scientists took as their model the replication of human functions through technological means. Their achievements pointed the way toward solving most of the practical problems of living. Apart from maintaining ourselves and working on such problems, we continued to ask whether there is some meaningful role for humans? Is there a role that is both important and necessary?

Naturski: (Nonette paused less between sentences than usual. Looking at Fortran plugged into the wall she began.) My batteries are running down too. Or perhaps humanity's batteries are running down. It is late in the evening for us and for all of humanity too. Or if we want to be optimists, it is early in the morning for a better than human future. A nice glass of cold water will revive me. Would anyone like one? (All nodded in agreement. Nonette left to get the water.)

Geist: (Becket was more emotionally wound-up than usual.) There you have it. We are needed to play the fallback role. When Fortran retires for the night, what needs to happen? His functions need to be assumed by us. From the point of view of his systems, we are part of his redundancy. He has us as an extra arm so to speak.

Abducto: (Wilfrid spoke overly deliberately as if his thought required effort.) On the other hand, without us, no one would need to go get water.

Geist: Our needs supply the rationale for our systems and Fortran's systems. Don't they? Fortran's needs are a residue of certain engineering problems. He needs to have his battery pack recharged. The engineers could just as well have given him a long cord.

Abducto: But with bioengineering aren't we being engineered so that some of our needs are a residual effect of an engineering problem too? The man with the Proto 2000 artificial heart has to have his batteries replaced periodically.

Naturski: (She overheard the conversation.) Yes but the needs with which we begin are human needs. (She brought in the water.)

Geist: Technology, or Nonette, achieves the satisfaction of most of them! (He held up the glass of water and looked to see if we understood his wry comment.)

Abducto: The deficits in the human design problem will have been solved!

Naturski: The problem is that our means for solving them has needs too. This gives us an additional role as machine minders. At least as far as Fortran is concerned, we are indispensable as redundant systems for fulfilling his needs. Our mutual dependency with machines assures that neither of us is obsolete.

Geist: Given that some engineer takes a human power, like memory, and builds a memory unit with much more of that power. Another takes the ability to do arithmetic and builds a calculator that is able to far surpass human performance in doing arithmetic. We repeat that process ad nauseam. Don't we? (He paused only long enough so that no one had time to reply.) There is now a whole constellation of tools that perform certain functions better than we do. As has often been said, we do nothing best. Our collective strength lies in doing many things well. (Becket spoke as if he were reminding us of one of his well-worn themes.)

Abducto: This is how we turned ourselves into redundant backup systems for the constellation of machines.

Naturski: Perhaps we are missing the forest for the trees. What we do best is build a battery of support technologies and maintain them. We are custodians, or janitors, in the tech world.

Geist: I think that is a very accurate statement of the history of the matter.

Abducto: I agree Becket, but Nonette, machines have taken over much of that role too. As you reported from an earlier conversation, machines are used to build machines. Machines are used to maintain machines. The axiom of the age of intelligent machines is that for every human function, there is a machine that could perform that function better than we can.

Naturski: Any counter example to your axiom Wilfrid can serve as a challenge to engineers to build such a machine. Since machines that replicate such functions are at least possible, probably feasible, and most likely already built or being built, it would appear that your axiom is irrefutable.

Geist: I disagree. Subjective experience can't be replicated in any but subjective beings. The experience of crying tears, having a belly laugh, enjoying the smell of coffee brewing, and appreciating the design of an oak leaf specify processes that arise in only some biological beings. There is no sign that any machine can respond in such a way. Oh sure, we can program a machine to say, "Ha! Ha!" and have its viscera heave up and down. This is no experience of laughter.

Naturski: I would think that if it were an experience of laughter, then we would have a biological machine. As Fortran would say, "All we would need is a bio-mechanical device like yourselves."

Geist: So you are saying that once the full spectrum of beings emerges from the biotechnological revolution, we may find our turf is crowded with alternate types of beings that have subjective experience.

Naturski: This would reduce our uniqueness or specialness. Many of these beings could take over the redundancy function for our machines.

Geist: We are far away from such developments. While I don't see any reason why such developments are improbable in the long run, the proof would be in the realities in question. I concede that this is at least one possible future.

Abducto: During your exchange, it came to mind that maybe we are missing the desert for the sand— to avoid that other hackneyed metaphor. (He chuckled at his humor. Nonette and Becket rolled their eyes.) We can always suppose that some machine or another can replicate some function or another. We think this way because we think in terms of probabilities. These are based on inductions from past experience. Because replication of human performance has developed this far, it appears to be likely that other steps in this direction are likely too.

Naturski: Such inductions seem wide open.

Abducto: The impossible, however, is not probable. All we would need to show is that it is probable that it is impossible that something is the case, and we would have a good argument for constraining those inductions.

Geist: There are different senses of "impossible." If we take graphic impossibility, then the probability would be high that something is impossible. For example, let's take squaring the circle. It is highly probable that this is impossible.

Naturski: Trying to square the circle is not some function that is uniquely human. A machine can try to square the circle too. We as well as a machine can attempt to square the circle.

Geist: A machine can attempt to square the circle endlessly spinning its wheels on and on. We humans would not even make the attempt. We would hardly get started. We would see that it is impossible and not even try to do it.

Abducto: What other senses of "impossible" do you have in mind Becket?

Geist: Topical impossibility implies that a topic has not been thought of yet. A discussion on that topic would be impossible. In centuries prior to the last one, Carbon Sixty had not been discovered. It would have been impossible for Plato to discuss Carbon Sixty in one of his dialogs.

Naturski: We could then ask, "How probable is this alleged impossibility?" We would examine the historical record to see how likely it would be that Carbon Sixty could have been discussed directly or indirectly by persons during that period of time. For all that we know, for example, Carbon Sixty was a twentieth century discovery. German chemists of the nineteenth century did not know about it, but it would be more likely that a German chemist in 1890 could have discussed it than that Plato could have. This is because chemistry in nineteenth century Germany was highly advanced compared to ancient Greece.

Geist: Hence, it is probably impossible for a machine to discuss a topic that is based on future discoveries. Statements about the past do not entail that piece of information. An induction to that topic would require a whole network of claims for which there presently is no rationale. The topic would have to come "out of the blue" so to speak.

Abducto: As you were developing your example, a related form of impossibility came to mind. What about inquiry impossibility? Every inquiry is on a topic. If we can't think of a topic, then we can't inquire into it.

Naturski: We would not have enough of the conceptual context to ask a question. We would not have identified a referent, causes or effects of such a referent, concepts describing that referent, a term designating it, or a collection of concepts which would allow us to synthesize an understanding of it. (Said all with one breath.) Only through these forms of context or their absence could we project the likelihood of what is impossible.

Abducto: Exactly! We can apply your kinds of context to the conversations with Fortran. What were you doing with him? If we impartially examine your activities, we see the marks of human dialog. This is significant because as a tool, Fortran could be put to any task involving cold calculations or even scattered functions that are devoid of the appearance of a unified self. He could, for example, be hooked up to regulate the water treatment system in Detroit. In most of his functioning, he is just a piece of equipment. His engineers also designed him to be able to participate in human conversation. In conversing with him, you supplied the content that made his responses human. Fortran was playing the "straight man" to your routines.

What was your content? You were conducting a series of inquiries. In light of these conversations, if you ask me what dominant function did humans play, I would have to say inquiry. I would then project that role into the human future. What is there for humans to do? Conduct inquiries. What role do humans have in the future of machine intelligence? Humans supply the content for automata in order to further inquiries.

Naturski: Your point about inquiry contains a very interesting insight Wilfrid. We have trouble identifying that which we are almost constantly doing; inquiry is too commonplace.

Geist: In more ways than one. If you ask persons what they are doing, few would be able to step back from their activities to identify such a thing as inquiry. If we identify their activity as inquiring and ask them what they do when inquiring, they would probably say something like, "Ask questions." Through personal experience, I have found that persons giving this response are merely reporting a meaning of the word "inquire." They are not revealing what they actually do. I mean if we examine the person's inquiries, that is, their actual performance, we often find no interrogatories. This is a sign that persons have difficulty identifying what they do when inquiring and that most people have not thought about it very much.

Naturski: What you say conforms to my experience too. We generate inquiry upon inquiry without thinking about how we are doing it. Often we are lost for words when trying to discuss our active inquiring. We resort to talking about creativity, intuition, and inspiration. As with the ancient Greeks, we leave it to the muses to speak through us.

Geist: Inquiry is also a social activity. We let the muses speak and incorporate their revelations into our dialogs. We examine their speech using others as a sounding board or as co-inquirers. The entire process is exciting and ennobling. This is very much the romantic vision of the Socratic tradition.

Abducto: I don't want to rain on your parade, but not all inquiry is about ennobling subjects. It is often about gossipy, mundane, or starkly practical matters. What I find surprising is that we shift from this level of content to profound issues without thinking about it— without realizing that a significant shift has been made. This seems to indicate that the activity is second nature to us and that our doing it is not always driven by content factors. It is performed across contents without considering their subject. An activity so ubiquitous and second nature to us indicates that the human future, if we survive, will very much involve inquiry. Even if we don't survive because we deliberately choose to migrate into the cyber-world or transmute into another species, our successors probably will still conduct inquiries.

Naturski: If we take the evolutionary perspective, humans conducted inquiries for perhaps more than one hundred thousand years. In using speech, they continually sought to find out about things. If we could hear a Neolithic community in conversation without understanding their language, it would be the human version of the sort of chatter we hear in the wild— forest sounds. The capacity to inquire is central to social organization and we would expect that inquiries would be mixed with other speech acts. Over time it would be adapted to non-social uses. From unreflective social employments to self-reflective applications for the individual's deliberate purposes. It is but a short step from this to doing science. It is very similar in magnitude to the transition depicted in Stanley Kubrick's 2001 A Space Odyssey. The ape-people used bones as weapons. This led directly to space travel.[1]

Geist: The doing of inquiry seems to be the big step culturally. Eventually we have the individual genius breaking through well-worn patterns of thought and asking an "unthought of" question. Someone like Leonardo Da Vinci was able to dream independent of his time and place. He was able to conduct inquiries, ask questions that set precedents for further inquiries.

Naturski: Da Vinci desired to fly, but the desire to fly is within human history from at least the time of the legend of Icarus. Dreaming of things is an ennobling human capacity. The rarity of quality dreams like Da Vinci's, however, indicates that quality dreams are an unlikely human activity.

Abducto: Da Vinci did not just dream and talk about dreams. He did something very preposterous. He decided to engineer wings!

Naturski: The idea of doing that is not very special. In the legend, Icarus made wings of feathers and wax and flew until the sun melted the wax.

Abducto: In the fictional world of Icarus, we think that he may be able to fly that way. We are not disappointed. He does fly until the sun melts the wax. Adult readers of the legend, however, know right away that wax and feathers won't work. They know that in our world, wax and feathers are far too heavy even for gliding. It is probably impossible for humans, unaided, to fly like birds.

Geist: Da Vinci did not research flying. He did not study the myth of Icarus and draw conclusions from it. As you said Wilfrid, he set out to dream the dream and design the machine.

Abducto: Yes, I still think it preposterous that he decided to engineer wings. After reading the legend, we didn't see people trying to build better wings out of feathers and wax. Da Vinci treated flying like any other engineering problem. The question is not whether but how. He realized that humans unaided by technical means couldn't fly through the air like a bird. He realized that it is technically impossible to use conventional aids to fly. He tried to discover technical means that would work. He thought that the key to developing technical means is to closely study birds.

Naturski: It seems like we would need to conduct a thousand inquiries in order to design technical means for flying.

Abducto: And this is where experience teaches us about an activity. After hypotheses are formed about aspects of the problem, the "tinkerer" needs to test those hypotheses. We would need to build on what works.

Naturski: Your pragmatist colors are showing Wilfrid.

Geist: Pragmatism has its place. We wouldn't expect that Da Vinci would have a drawing of a Boeing 767 in his notebooks! That would be probably impossible! His project was to build a human-bird using machine technology of his day.

Naturski: He of course had no tradition of technical expertise in electronics, jet mechanics, fuel science, airfoils, and so on.

Abducto: The evolution of these sciences required sustained inquiry over much time by large numbers of people. It is this cultural infrastructure that develops, supports, and sustains these sciences. The infrastructure is responsible for contemporary technical revolutions. Da Vinci's dream could be only that.

Naturski: So the big challenge historically was not individual insight but the ability to keep inquiry going conceptually, procedurally, technically, materially, and socially. (Said without taking a breath.)

The lack of interest on the part of Da Vinci's patrons led him to be a kind of renaissance Nowhere Man. His notebooks as far as he knew were "nowhere plans for nobody."[2] Da Vinci's patrons were not interested in devoting their resources to flying.

Abducto: Hence it is probably impossible that cyborgs and biotechnological beings can leapfrog over us in inquiry. They are beholden to contingent developments, dependent on experience, and constrained by the devotion of cultural resources just as we are. And the only cultural resources that exist, that is, that are suitable for this purpose, are ours. Moreover, they would need large sub-cultures of scientific expertise that would allow them to transcend their social and material circumstances through inquiry. In effect, a machine culture of interacting machines would have to evolve.

Geist: Another necessary step on the road to transcendence is the development of philosophy. Raising questions of ultimate concern about the origin of things, the nature of value, and conceptions of truth is a precursor and foundation for myriad empirical inquiries. Your namesake (As he spoke to Abducto.) Wilfrid Sellars pointed out that terms in ordinary discourse, like "origin," "value," and "truth," are typical of pre-science and end up writ large in science.[3] We would not understand science very well without awareness of this debt. Much scientific language is drawn out of ordinary language as systematized by philosophical thinking.

Naturski: Outside of philosophy and science, philosophers propose that raising such questions is of value in itself. Bertrand Russell is right that these questions can't be answered in a final way, but he nonetheless thinks that each generation of humans should be encouraged to ask them and confront answers to them.[4] Each generation is thought to humanize itself through philosophical thought— to join the conversation begun in ancient times— to reflect upon themselves asking and trying to answer these questions. (All said in a tongue-in-cheek air.)

Geist: I detect a tone of disapproval Nonette. I agree that this view is all wrong. Layers of general theory are useful in rethinking scientific inquiries and in providing them with a context. Most of this theory comes from the philosophical tradition and not from the common person confronting philosophical questions afresh. Conceptual progress has been made with ultimate questions. For the common person to raise philosophical questions in their barren generality is a pointless exercise.

Abducto: Isn't your claim a Straw Person? Scientists are ordinary persons too. They use ordinary discourse apart from philosophy in expressing their basic prescientific understanding of things.

Geist: As a fact, many do. My argument is that they would be better served if many of these ideas were developed through examining others philosophical inquiries. My problem is with education. Should an education for all persons require philosophical inquiry about large questions that quickly reaches a dead end? I think that philosophical inquiry is not supposed to be a Zen exercise with no answer.

Naturski: It is not supposed to be but doesn't inquiry of this sort lead to much the same thing for all of us? Isn't reason defeated in trying to face such questions?

Geist: I don't think so. When a novice asks such questions, the result is usually a quick dead end. For the professional philosopher, the inquiry continues. It continues not as a person's private inquiry but an inquiry built upon a literature and prior discussions. This is what leads to conceptual progress. The same dead ends are not reached again and again. New dead ends spell conceptual progress, and no matter what dead ends are faced, philosophers continue the inquiry. (He laughed at how strange this sounded.) This means that there is much to explore, much that has not been exhausted.

Abducto: You seem to have a point although I have trouble telling whether you are taking yourself seriously. Whether we think there is much that is interesting being said or not, it is undeniable that historically important developments continue to occur. Isn't that because fundamental philosophical questions are so general that they stand for an endless number of different questions? For example, if we ask the primary metaphysical question, "What is there?" we can go on and on asking about and talking about just about anything.

Geist: In inquiry terms, a fundamental question is the source of a large number of sub-questions. Each sub-question provides part of an answer to the fundamental question. In this way, the generation of sub-questions is constrained. What determines whether a sub-question does provide part of an answer to a fundamental question? We consult the semantic resources of the question. We explore its referents. We understand how it is related to bodies of theory. We examine the relevance of its potential answers. We determine what can't count as its answers.

Naturski: At this time in the morning I don't intend to be irritating, but can't some ordinary persons apply your criteria and make headway? The criteria seem to be no more than "advanced" common sense.

Geist: They are "advanced" because they are gleaned from the philosophical tradition. In order not to reinvent the wheel, philosophical inquiries utilizing such criteria are informed by the tradition. We understand its precedents, logic, methods, and language.

Abducto: By the time we assemble the history of inquiry, train in its methods, and apply your sophisticated criteria, we are professional philosophers engaged in the endless exercise of trying to answer fundamental questions. They are a bottomless source of philosophical activity.

Geist: The scope of the inquiries indicates that there is no final solution that is simple. Given the complex concepts that frame the inquiry's questions, however, we would expect as much. For each sub-question, we can raise other sub-questions and other sub-questions under those. In such a hierarchy, questions at the bottom would be more digestible than those higher up.

Abducto: Considered this way, philosophy certainly has a future. (Said in ironical tone.)

Naturski: The nature of philosophical inquiry is very similar to the nature of scientific inquiry, so education in both would be similar.

Abducto: Perhaps because they were riding the same historical track for more than two millennia.

Geist: The humanization of the young through philosophy, then, would require introduction to philosophical inquiry through its tradition, an understanding of inquiry into philosophical questions, and a view of how that process is open-ended but yet constrained.

Abducto: I will agree with you so long as sub-questions are well formed. They should be more explicit than fundamental questions and be better focused. With our eye on the fundamental question, we are ennobled through that special kind of self-reflection. On the other hand, a paradox arises with your view. Participating in this humanizing influence depends upon not reinventing the wheel, that is, by conducting new inquiries. Must they be new to humanity at large?

Naturski: Inquiries that have been resolved need not be re-conducted. If they were, then there would be a flaw in the educational system. The transfer of culture to succeeding generations would not be efficient. Put this way, the task for educators is daunting. On this measure, even the best education is largely a failure.

Abducto: Need the inquiry be new to humanity at large? What would an educational system be like which required such inquiries? I believe the issue lies in how young people would come to conduct inquiries new to humanity. What would this presuppose? (A series of answers quickly came to mind. Geist began.)

Geist: Sometimes it is said that the "how" of inquiry is more important than the "what." At least insofar as learning to inquire is concerned, we would not need to be unduly concerned that some person had followed a line of inquiry before. In fact, some classic inquiries can serve good pedagogical purpose as effective means for rehearsing inquiry skills.

Abducto: Secondly, I think that it is a mistake to view inquiry in an atomistic way. Inquiries often come nested like Russian matryoshka dolls. Each layer expresses topics of wider scope. From this point of view, in order to conduct a new inquiry, we would need to extend a prior inquiry. In order to begin we would need to work up to the point where the prior inquiry left off. We would attach narrower or more specific questions to, let us say, an inquiry of medium scope.

Naturski: Part of the context of inquiry is the point of view taken. Each point of view would initiate a different inquiry into the same topic. For example, we could conduct Kepler's inquiries into the movements of the planets using Newton's point of view on gravity. We may end up with a view largely similar to Kepler's but the inquiry leading to that view would be different.

Geist: I heard three reasons, so fourthly, there is the daunting historical question. We have scant evidence of inquiry's past. I mean the events comprising process and product of most inquiries is lost in the passing of the private lives of individuals. The millions of inquiries conducted by each of us in the course of a lifetime is hardly noticed let alone taken to be important enough to record.

Abducto: Even in disciplines within the sciences and humanities, the products of inquiries are seldom written up.

Naturski: How useful would it be to do so? Is there some psychological interest in finding out about actual performance? The doing of inquiry, however, seems to surpass the history of doing it on the level of the common individual.

Abducto: In order to do it better, we should be able to see detailed examples of how it is done. We could record examples of performance and analyze them. This would remove some of the failure that results from trial and error. Why reinvent the wheel in this context?

Geist: We would not want, however, for the examples to be treated as norms that would crimp and curtail further inquiry.

Abducto: We would want to maintain maximum efficaciousness.

Naturski: From this exchange, we can suggest ways in which the educational system needs to be improved. If inquiry is as important as we are taking it to be, young people should be schooled in its history as well as in its practice. Examples of performance should be supplemented by analysis of detail.

Abducto: A certain measure of such curricula is only prudent. A contrarian, however, could make a point against your proposal by appealing to parsimony. Inquiries are integrated with disciplines. Every theoretical science aims toward contributing to knowledge in that science. Inquiry as activity is the means to achieving that goal. Successful inquiry is supposed to end with knowledge. Thus, if students study a science by learning to conduct inquiries within it, they will learn all they need to know about inquiry in that area. Multiply these efforts with each new subject studied, and students will have mastered inquiry without studying it as an independent subject. It would be a mistake to remove inquiry from its pragmatic context.

Naturski: Here is a quick reply Wilfrid. The history, psychology, and sociology of inquiry in a science are not ordinarily treated in the daily doing of that science. We need to take a step back from the heat of inquiry to examine it from other points of view. This leads into what could be extended into a longer reply. The examination of inquiry in general will reveal some broad principles, models, techniques, strategies, and analytical concepts that cut across disciplines and are worth learning on their own. There will be enough material requiring special treatment so that an independent subject will be born.

Geist: A second contrarian reply stems from one of Fortran's arguments for human obsolescence. Machines will conduct our inquiries for us. Cyborgs will save us that labor. Moreover, cyborgs will do that job more efficiently. Thus, few humans need to be schooled in inquiry.

Abducto: But have we not just indicated presuppositions that machines presently can't fulfill? It is probably impossible that they will do so in the near future.

Geist: But with our support, it is probable that machines like Fortran will continue to advance as before. Over much time, our role will gradually diminish.

Naturski: Even if machines did a better job inquiring than us, we would still profit from learning to inquire in school. I will use an argument from analogy. Suppose we invent a scent detector that works better than the nose of a bloodhound. We could argue that bloodhounds are obsolete. We don't need to put them on someone's trail. If we look at the matter from the point of view of the bloodhound, however, the animal should follow scents. It is part of the animal's nature. The animal has the proper equipment, enjoys following scents, achieves some degree of fulfillment following them, and would follow scents without our encouragement. Likewise, we humans inquire as part of our nature. We have the mind and temperament for it. We enjoy it. We achieve some degree of fulfillment doing it. We will inquire even in absence of serious societal goals. Education should prepare us for meaningful lives. Learning to inquire and about inquiry is an important step toward that end.

Abducto: Tigers are equipped to tear apart prey. By the fact that such behavior bursts upon the scene, we can view it as fulfilling. If tigers enjoy anything, they enjoy tearing apart prey. Should we give tigers in zoos live animals to feed on? No. We should give them chunks of horse- meat instead. In the zoo life, some tiger instincts are no longer applicable. In this context, such instincts are obsolete. Likewise, some human instincts become obsolete in the cyber world.

Naturski: The reason we don't feed tigers live animals is for ethical reasons. We don't want to be cruel and inhumane to some animals in order that others are fulfilled. If that would be the only way that tigers would eat, we would have to feed them live prey in order to keep them in zoos.

Abducto: There are other ways to fulfill humans than to have them do inquiry.

Naturski: But there are no ethical objections to humans doing inquiry. Having some person ask questions and pursue answers to them is harming no one.

Abducto: That is where you are wrong Nonette. Inquiry as a recommendation for humanity in the large is not a parlor entertainment. Meaningful inquiry requires resources and social organization. It would be a political decision whether society's resources should be so invested. Society decides, "Would it advance social utility to allow humans to fulfill themselves in this way?"

Naturski: It is also a political decision as to whether we should leave all meaningful inquiry to cyborgs. So what if it takes us a bit longer to reach adequate results? The time lag sometimes means that we have built up the necessary context for further inquiry.

Geist: You are pointing out that machine efficiency is just one value. Other human values may be more important or at least should be given some priority in decision-making. Take human fulfillment and human flourishing. They are important values too. We still have much to learn from Aristotle on this subject.

Abducto: Progress in inquiry is another significant value. I would make another argument in favor of human inquiry Nonette. No matter what competency a machine has, it will always be one step behind the human collectivity. Tools are designed for purposes— to perform certain functions. A different set of purposes would produce a different tool. If tools follow purposes and machines are tools, then the design of machines follows purposes. So, even if a large part of the labor in inquiry is done by machine, the raison detre of the machine is subordinate to purposes related to the inquiry.

Naturski: Once we have a tool, however, we find other purposes for it. Our technical expertise in a certain area often drives inquiry. Thus, regardless of the original purpose for a tool, Cyborgs could assemble arrays of machines to perform new tasks.

Geist: Furthermore, cyborgs can detect gaps in machine expertise. Suppose that Fortran does function X, and Cobol does function Z. Both Fortran and Cobol could pick up the gap, that is function Y, and design a machine called Pascal to fill that gap. Once Pascal is designed, they instruct us how to proceed in building it. I am not convinced, Wilfrid, that we will have a robust role in devising new machines.

Abducto: Let's grant your points and further grant that machines have self-generated purposes of their own in inquiry. Until machines convince us that needed or strategically focused inquiries will be conducted by them, we will proceed to conduct our own inquiries usually with their aid.

Geist: I agree Wilfrid. Suppose that a machine could be designed to serve any inquiry purpose whatsoever. Of the infinity of inquiries that could be conducted, which should be conducted next? To answer this question, the machine would have to perform at least at the level of a team of competent research scientists.

Naturski: I think that you are granting too much to machine capabilities. If machines could serve any inquiry purpose whatsoever, it would have to be a universal inquiry machine. It could be a configuration of all the cyborgs reprogrammed and linked together. Would it be universal? If so, the "how" of inquiry would be resolved and the machine would be waiting for us to supply the "what." To the contrary, at this time I don't think we have the foggiest notion of how to bring closure to the "how" of inquiry.

Abducto: The "how" of inquiry is set up by questions. Until we bring closure to questions, we cannot bring closure to finding answers to them. In fact, new types of questions would show that the machine is incomplete.

Geist: New types of questions are always possible. Individuals of genius often think of new ones. Kant's question, "How is experience possible?" gave rise to a whole host of questions about possibility.

Abducto: We already knew how to answer Kant's question. It is a "how-question." We would find a way to explain experience. You know that it would be abduction to the best explanation. (He said with a twinkle in his eyes.)

Naturski: If you are saying that content doesn't matter, then the possible types of questions are already known. They are either "response" questions or wh-questions. Response questions require an affirmative or negative response. For example, "Did G.E. Moore write Principia Ethica?" "Yes, he did." "Did Immanuel Kant write A Treatise of Human Nature?" "No, he didn't." By contrast, wh-questions employ the interrogatory pronouns and adverbs: who, what, where, when, why, how, and which.

Geist: You can add "whence, whom, and whose." Also add compound words such as whatever, whoever, whereof, and whereto.

Abducto: I did not think of those Becket. I see that both of you have thought about this subject before. Let me add that our list of wh-words comprises types of questions. I mean that inquiries are typed by saying, "Oh! That is a why question," or "That is a what question." They are our tools in inquiry. Inquiry is flourishing in this century. In particular, millions of formal inquiries have been conducted within institutionalized science. The brute number of people with advanced educations suggests masses of people conducting complex and sophisticated inquiries as never before. Yet the number of wh-words is stable. This may indicate that each wh-word interrogates a fundamental category as Aristotle suggested.

Geist: If you will pardon my abruptness, I think that view is false. Their number is not stable. The history of English reveals a number of archaic wh-words. We have wherefore, whereunto, wheresoever, whither, whithersoever, and whitherward. (He smiled.) Hence there were more wh-words in use in the recent past when less formal inquiring was going on than now when more formal inquiring is going on. This state of affairs is not what I would have expected either Wilfrid.

Naturski: It may be that science relies on using some wh-questions over and over.

Abducto: Perhaps scientific inquiry just indicates the richness of a certain core group of types of questions.

Naturski: Besides how would "wheresoever" or "whitherward" significantly extend scientific inquiry? How would such questions be a part of scientific inquiry at all?

Geist: Lord only knows. I think you are right Nonette that the maximum vocabulary does not necessarily lead to the most efficacious use. In fact, scientific inquiry may be flourishing independent of wh-questions. We have alternate linguistic means for performing inquiry functions. For example, Katz and Postal proposed using imperatives.[5] Instead of asking, "Who hit the most home runs in a single season?" we can say, "Tell me the name of the person who hit the most home runs in a single season." The advantage of this analysis is that there are far more verbs that can be used to give instructions than there are wh-words. For this reason, instructions or requests for information can be more articulate, precise, and less context dependent than questions.

Naturski: On the downside, instructions are directed to someone. Where no specific person is identified, the intention is left open to any person. For example, "First, sift the flour," is directed to anyone reading the recipe. But, imperatives have to make sense as instructions. Many questions, by contrast, are person-independent. "Why do brains generate thought?" asks about brains and does not tell anyone to do anything. It does not necessarily imply, "Tell me why brains generate thought."

Abducto: Yes, asking about a phenomenon is different than requesting that someone explain something about a phenomenon. "Why is there a solar wind?" is timeless. It asks no one in particular to answer it. Different researchers can address pieces of the question. It is open as to whether there is an explanation for the solar wind. It is a rational question even if no one can presently answer it. "Explain the solar wind," is a request for an explanation. It is time bound. Someone is supposed to take it as an instruction. It presupposes that the instruction can be followed. It would not be a rational instruction if no one can presently follow it.

Naturski: Bravo Abducto! I like your points of contrast. I might add, to the contrary, that instructions can be somewhat timeless, person independent, open, and rational even if they can't be presently followed. But this is because in our language imperatives are becoming substitutes for questions by having them assume the semantics of questions. This masks the differences that you pointed out. Nonetheless, the differences are there.

Abducto: (Wilfrid looked troubled.) What troubles me is the expansion of the number of wh-words. Given Becket's point that the number in use has dwindled, from reverse perspective, can there be many new ones for us to discover or invent? The impression we have is that the basic wh-words were always part of the language and any other language that preceded it. Are there such categories of thought?

Geist: It shouldn't trouble you Wilfrid. You are a champion of human inquiry. More wh-words would expand possibilities for inquiry. Suppose we think of new wh-words that have the significance and importance of a "why" or a "how." A new day would dawn for the human mind. Our inquiries would be directed toward undreamed of answers. Moreover, if we could understand the principles for formulating new wh-words that are of the significance of "why" and "how," we would open the human mind to all things in any dimension.

Abducto: What if our current wh-words are few and insignificant compared to those possible? Our inquiries may be provincial. We may have little understanding of our limitations.

Naturski: All the better for the human future. I share Becket's optimism. The cyborgs can have our interrogatories. We are the inventors of new ones!

Abducto: My question is, "How would we go about thinking of new wh-words?" If this job is so easy, why hasn't anyone done it before?

Geist: I think we need to look to Aristotle's discussion of predication in relation to categories. I recall a fertile passage that we can use to begin our inquiry. (Becket removed a copy of Aristotle's <u>Topics</u> from the bookshelf.) He says,

> "… the man who signifies what something is signifies sometimes a substance, sometimes a quality, sometimes some one of the other types of predicate. For when a man is set before him and he says that what is set there is a man or an animal, he states what it is and signifies a substance; but when a white colour is set before him and he says that what is set there is white or is a colour, he states what it is and signifies a quality. Likewise, also, if a magnitude of a cubit be set before him and he says that what is set there is a cubit or a magnitude, he will be describing what it is and signifying a quantity. Likewise, also, in other cases; for each of these kinds of predicate, …."[6]

He is saying that "What?" may ask about a substance, quality or quantity. The answer to the question would give or specify a substance, quality, or quantity.

Suppose that we just asked, "What?" We would not know whether someone was asking about a substance, quality, or quantity. Context should reveal which category a person was asking about. The other way we could proceed, however, is just to ask, "What substance?" or "What quality?" or "What quantity?" The "what" asks us to designate or specify a member of that category. Other wh-words function this way only with different categories. In other words, a what-question can be given for other wh-questions.

Naturski: I am not following you Becket.

Geist: The idea is that we can ask about other categories. For example, "Who?" asks, "What person?" "Which?" says, "What member?"

Naturski: So then, "When?" asks "What time?" "Where" expresses, "What location?"

Abducto: The big question "Why?" can be taken as, "What reason, cause, or purpose?" "How?" can be construed as, "What manner or sequence?"

Geist: We have just demonstrated that wh-questions are or are types of what-questions.

Abducto: In English, the wh-words reveal to us the categories that have become conventional such as time, manner, reason or cause.

Naturski: So what-questions corresponding to the wh-words don't reveal new categories of thought. The topics of person, location, time, and reason existed before we asked what about them.

Abducto: If I understand you correctly, you are saying that the conceptions of reason, cause, and purpose existed prior to the word "why."

Geist: I believe that "why" originated in Old English as the instrumental case of "what." Prior to this period, no human being asked, "Why?"

Abducto: That is preposterous. The question came up in ancient times. Plato gives responses to why-questions very commonly.

Naturski: I think that Becket is teasing you Wilfrid.

Geist: People asked, "Why?" without using the word "why." They asked using other words like, what reason, cause, or purpose. For example, Plato used the categories such as logos and telos.

Naturski: All of those discussions about the magic of asking why in philosophy are to some degree misguided. (In pseudo-serious tone, she proceeded.) Can we do philosophy without asking why? Can we do inquiry without asking why-questions? What would human intellectual life be like without being able to ask why? The answer is simple. All will be the same because we could ask, "What reason?" or "What cause?" or "What purpose?"

Abducto: If Aristotle is right, then, "What?" is the fundamental question. It can be used to ask for items in any of the most basic categories. So then, other wh-words signify sorts of what-questions. If we want to invent new wh-words, we would only have to invent new what-phrases and coin tokens to stand for them. Let's see if we can do that.

Geist: I was always troubled by personal pronouns for persons only. Our language has built-in biases against other animals. Often other animals are designated as mere things. To rectify this problem, just as "Who?" asks for a person, I propose that "Whe?" asks for a non-human animal. "Who is that? That is Jack." "Whe is that? That is Fido."

Naturski: And "whe" would function as a relative pronoun. "That is the person who ate the pie." "That is the animal whe ate the table scraps."

Geist: With the marking off of the category of animals, they would be distinguished from humans and inanimate things. When asking, "Whe?" we know that the answer would be the specific animal or "the so and so." In other words, "Fido" or "the dog."

Abducto: Your example brings another to mind. In many recent philosophical discussions, inquirers are supposed to be sensitive to the context of a subject. No wh-word asks specifically, "What context?" We could coin "Whix?" for this purpose. "Whix frames the issue?" "Whix" could be defined as "an interrogatory used to ask for contextual background especially as with an issue."

Naturski: I have another example. Statistical language has expanded greatly, and we have no wh-word that asks for representative samples. "Whar?" can mean, "What representative sample?" For instance, "Whar profiles women over fifty years old who are candidates for estrogen therapy?" It would be beneficial for researchers to identify some topic as a "whar question".

Abducto: A logical example is that of an open disjunction. At times we wish to acknowledge that a disjunction does not reveal all alternatives. Sometimes we wish to begin at that point. By asking the question, "Whu?" the inquirer would only be looking for a portion of the disjunction. "Whu are suspects in the assassination of John F. Kennedy?" "They are the Soviets, Castro, Cubans, CIA, the mob, Oswald, Lyndon Johnson, and/or so forth."

Geist: "Whok?" can mean, "What computer program?" "Whok has the best graphical interface?" Ah, "What computer program has the best graphical interface?" is a whok question rather than a whu question! (He laughed.)

Naturski: I suppose that we could go on this way. It is surprising that we don't have a large vocabulary of wh-words.

Geist: Wouldn't such a vocabulary be merely for convenience? Would it be needed?

Abducto: (Wilfrid's look gave him the floor, but he paused before he spoke.) Some other pragmatic considerations come to mind. New wh-words can replace what-phrases that are used only in passing with what will become habitually employed types of inquiries. Their conventional standing would command a share of the life of the mind. Secondly, conventional categories of inquiry allow for distinguishing types of queries from others. You (He looked at Becket.) just did this with whok and whu. I did it before by saying, "That is a what-question, and this is a why-question." Thirdly, wh-words such as "why" have a venerable history where expansive bodies of meaning have become associated with them. As part of that history, the theory of reasons, causes, and purposes has developed widely in the modern period. Some of this theory has become incorporated in the sense of "why" as a simple interrogatory.

Naturski: But since "why" is a what-derivative, we would expect this theory to be associated with "what." It is not associated with "what" by itself. Some of it may be associated with a "what phrase" such as "What purpose...?"

Abducto: Thus, "what" by itself is free of the theory that accompanies "why".

Naturski: Following the metaphor of Willard Quine's web of belief, we can think about a web of inquiry with nodes as interrogatories. It is no small matter, then, which wh-words become conventional.

Abducto: I can't even imagine how cyborgs could be programmed to decide what could count as an efficacious node. Setting the parameters of these sorts of dispositions is quite opaque.

Geist: Our inquiry took us down the path of wh-words as what-derivatives. But what about "what" as a wh-word? By itself, "what?" can interrogate the category of categories. It could ask, "What category?" and under that question we would have categories such as person, place, or reason. They would be queried by the derivatives who, where, and why. But, besides what, "Could there be a wh-interrogatory which is not a what-derivative?"

Nonette: Do you mean a wh-interrogatory that asks for a new category? A category that is in addition to current categories?

Geist: Then it would simply ask, "What new category does something belong to?" Our examples of "whe" or "whok" did that job. They set up new categories. I am trying to ask about a wh-interrogatory that transcends categories. It would ask for something non-categorical or something incapable of being in a category. Can we get beyond categories?

Abducto: Aristotle would say no because there is nothing for it to interrogate. Let's see what results from trying to get beyond categories. For starters, we could easily create a token for the wh-word. Let's call it "Whef?" Unlike other wh-questions, "Whef?" would ask for nothing categorical. What could this mean? Could it ask for a particular that is not of a type?

Geist: But all things that are not of a type form the category of all such things! They would be of that type! The type called "the typeless."

Nonette: But then there would be no members of that type; it would be the null set. It cannot be a member of that set. If it belongs to that set, then it is of a type. If it is of a type, then it is not typeless. So, to be a member of that set, it would have to be typeless and not typeless at the same time.

Abducto: I see the contradiction Nonette. I think the problem to begin with is with calling the object of inquiry "typeless". Once we identify the object by using an identifier, then we can ask for the object by using that identifier. It would be of a current category or a new one. If we insist that "Whef?" transcends categories, this presents a conundrum. We shouldn't way, "<u>What</u> is 'Whef?' asking for. "<u>What</u>" leaves us in the realm of categories. "Whef" leaves us in the realm of whatever is non-categorical. We can ask, "'Whef?' asks for?" But here, are we beyond sense? Does whef ask that of which we cannot speak? Can "Whef?" ask at all? Is it disqualified from being a wh-question?

Narrator: The end-point of the dialog was reached. Maybe Wilfrid was falling into incoherence. By this time of the morning we were like the walking wounded. The exhilaration of the discussion kept us tied into our subjects but we were emotionally operating on fumes so to speak. With these last perplexing topics, the seven dialogs about the human future came to a close.

Given what has transpired since that time, I thought it important to establish a record of these conversations. My intent for transcribing them was to give us pause and underscore the imperative, the ethical imperative, of having conversations about the human future. Deliberative choice should be brought to bear on the emerging course of events. From a global perspective, the momentum of these transformative events is almost inevitable. In the individual case, we may mitigate their effects. We can at least try to mitigate their effects for us and for those in our small circle of family and friends.

[1] Stanley Kubrick, director. 2001 A Space Odyssey, Screenplay by Stanley Kubrick and Arthur C. Clarke, Metro-Goldwyn-Mayer, 1968.

[2] John Lennon, "Nowhere Man," The Beatles Illustrated Lyrics, Allan Aldridge, ed., Delacorte Press, N.Y., 1969, pg. 60.

[3] Wilfrid Sellars. Empiricism and the Philosophy of Mind, Harvard U. Press, Cambridge, Mass., 1997, pg. 81.

[4] Bertrand Russell. The Problems of Philosophy, Oxford University Press, New York, 1959, pgs. 154-58.

[5] Jerrold J. Katz and Paul M. Postal. An Integrated Theory of Linguistic Descriptions, MIT Press, Cambridge, Mass., 1964, p. 85.

[6] Aristotle. "Topics," in The Complete Works of Aristotle, W.A. Pickard-Cambridge, trans., in Jonathan Barnes, ed., Vol. I, Princeton University Press, Princeton, N.J.,1984, pgs. 172-173.

www.ingramcontent.com/pod-product-compliance
Lightning Source LLC
Chambersburg PA
CBHW052145070326
40689CB00050B/2091

* 9 7 8 1 4 7 8 7 1 0 1 8 9 *